No Nonser
Class License Study Guide

for tests given between July 2018 and June 2022

Dan Romanchik KB6NU

ISBN-13: 978-0-9832216-3-0

Copyright © 2018 Daniel M. Romanchik

All rights reserved. No part of this publication may be reproduced, stored in a retrieval system, or transmitted in any form or by any means, electronic, mechanical, recording or otherwise, without the prior written permission of the author.

CONTENTS

1 What is amateur (ham) radio?

5 Electrical principles

23 Electronic components and circuit diagrams

39 Radio wave characteristics

51 Antennas and feed lines

63 Amateur radio signals

71 Electrical safety

81 Amateur radio practices and station setup

89 Station equipment

99 Operating procedures

119 Rules and regulations

143 About the author

What is amateur (ham) radio?

Amateur radio, also known as ham radio, is a hobby enjoyed by hundreds of thousands of Americans and millions around the world. They enjoy communicating with one another via two-way radios and experimenting with antennas and electronic circuits.

All kinds of people are amateur radio operators, also known as "hams." Hams are young, old, men, women, boys and girls. Kids as young as seven years old have gotten amateur radio licenses, and many hams are active into their eighties and beyond. You never know who you'll run into on the amateur radio bands: young and old, teachers and students, engineers and scientists, doctors and nurses, mechanics and technicians, kings and entertainers.

For example, did you know that most of the astronauts sent up to the International Space Station (ISS) in the last five to ten years have been licensed radio amateurs? They use the amateur radio station on board the ISS to communicate with school groups all over the world as they are flying over.

How do you get into amateur radio?

With just a little study, you can learn all you need to know to get a Technician Class license, which is the license class designed for beginners. To get a Technician Class license, you must take a test with 35 multiple-choice questions and answer 26 questions correctly. The test covers basic regulations, operating practices, and electrical and electronics theory.

Knowing Morse Code is no longer required to get this license, nor any class of license. Technician Class licensees have all amateur radio privileges above 30 MHz, including the very popular 2-meter band. Technicians can also operate Morse Code (CW) on portions of the 80m, 40m, 15m, and 10m bands, and voice and digital modes on portions of the 10m band.

There are two other license classes: the General Class license and the Amateur Extra Class license. To get a General Class license, you must pass another 35-question test; the Amateur Extra Class test has 50 questions. The tests are progressively more difficult.

General Class licensees get phone and digital mode privileges on portions of the 160m, 80m, 60m, 40m, 20m, 17m, 15m, 12m, and 10m bands. They can also operate CW and digital modes on the 30m band. Amateur Extra licensees have all amateur privileges.

How much does it cost?

Basic study materials, such as this study guide, can be had for free, and the license exam fee will be $15 or less. Once you have your first license, most hams find it best to start with simple equipment and grow over time. A handheld VHF FM transceiver can be purchased for less than $100 new, and excellent used equipment is often available at low prices. All things considered, the cost to get the first license and radio should be less than $200.

Where do I take the test?

Amateur radio license examinations are given by Volunteer Examiners, or VEs. VEs are licensed radio amateurs who have been trained to administer amateur radio tests. To find out when the VEs in your area will be giving the test, go to the American Radio Relay League's (ARRL) website: http://www.arrl.org/find-an-amateur-radio-license-exam-session. Using that page, you will be able to search for test sessions that are close to you. If you do not have access to the Internet, you can phone the ARRL at 860-594-0200.

Can I really learn how to be an amateur radio operator from a study guide like this?

Yes and no. This manual will help you get your license, but getting your license is only the beginning. There is still much to learn, and to get the most out of amateur radio, you will have to continually learn new things.

This study guide will teach you the answers to the test questions, but will not give you a deep understanding of electronics, radio, or the rules and regulations. That will be up to you after you get your license.

I hope that, by helping you get your license, this guide will encourage you to become an active radio amateur and get on the air, participate in public service and emergency communications, join an amateur radio club, and experiment with radios, antennas, and circuits. These are the activities that will really help you learn about radio in depth, and in the end, help you be confident in your abilities as an amateur radio operator.

How do I use this study guide?

First, read through the study guide and then take some practice tests. The characters in parentheses —(T5A05), for example—refer to the question number in the Technician Class Exam Question Pool. You will find the answers to questions in bold. You can take practice tests by going to the following websites:

- AA9PW.com
- QRZ.com/hamtest/
- eHam.net/exams/ (http://eham.net/exams/)
- HamExam.org (http://hamexam.org)
- HamStudy.org (http://hamstudy.org)

There are also ham test apps for both iOS and Android tablets:

- iOS:
    - Amateur Radio Exam Prep (https://itunes.apple.com/us/app/amateur-radio-exam- prep-technician/id297951496?mt=8). $4.99
    - Ham Radio Exam (https://itunes.apple.com/us/app/ham-radio-exam- tech/id601991935?mt=8). FREE.
- Android:
    - Ham Radio Study (https://play.google.com/store/apps/details?id=com.tango11.hamstudy)
    - Ham Test Prep (https://play.google.com/store/apps/details?id=com.iversoft.ham.test.prep&hl=en)

Many of the questions use acronyms with which you may be unfamiliar. In the glossary, you will find definitions of those acronyms. Please refer to the glossary if you are unsure of the meaning of an acronym.

Good luck and have fun

I hope that you find this study guide useful and that you'll become a radio amateur. Remember that getting your license is just a start and that you will continue to learn new things.

If you have any comments, questions, compliments or complaints, I want to hear from you. E-mail me at cwgeek@kb6nu.com. My goal is to continually refine and improve this study guide.

Dan Romanchik KB6NU

Electrical principles

Units and terms: current, voltage, and resistance; alternating and direct current; conductors and insulators

Figure 1 shows a simple electric circuit. It consists of a voltage source (in this case a battery, labeled E), a resistor (labeled R), and some wires to connect the battery to the resistor. When connected in this way, the battery will cause a current (labeled I) to flow through the circuit.

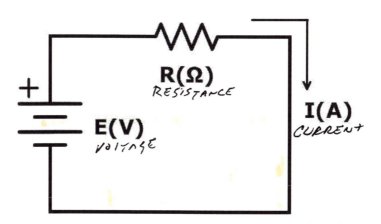

Figure 1. A simple electric circuit

The three basic parameters of this circuit are electromotive force (E), current (I), and resistance (R). Electromotive force, or EMF, is the force that causes electrons to flow in a circuit. We use the letter E to denote electromotive force. Electromotive force is measured in volts, and we use the letter V to denote volts.

> QUESTION: What is the electrical term for the electromotive force (EMF) that causes electron flow? (T5A05)
>
> ANSWER: **Voltage**
>
> QUESTION: What is the unit of electromotive force? (T5A11)
>
> ANSWER: **The volt**
>
> QUESTION: How much voltage does a mobile transceiver typically require? (T5A06)
>
> ANSWER: **About 12 volts**

Current is the flow of electrons in a circuit. In Figure 1, the letter I stands for current. Current flows from the positive (+) terminal of the voltage source through the circuit to the negative terminal of the voltage source. Current is measured in amperes, and we use the letter A to stand for amperes.

> QUESTION: What is the name for the flow of electrons in an electric circuit? (T5A03)
>
> ANSWER: **Current**
>
> QUESTION: Electrical current is measured in which of the following units? (T5A01)
>
> ANSWER: **Amperes**

Because the polarity of the battery voltage never changes, the current will flow in only one direction through the circuit. We call this direct current, or DC.

> QUESTION: What is the name for a current that flows only in one direction? (T5A04)
>
> ANSWER: **Direct current**

Batteries supply direct current, or simply, DC.

Resistance is the third parameter. As the name implies, resistance opposes the flow of electrons in a circuit. The higher the resistance, the smaller the current. We use the letter R to stand for resistance. Resistance is measured in ohms, and we use the Greek letter omega (Ω) to stand for ohms.

The type of current you get out of a wall socket is different from the current that you get from a battery. Unlike the battery, the polarity of the voltage changes from positive to negative and back to positive on a regular basis. In fact, it changes polarity 120 times per second. This means that the current changes direction 120 times per second. Because of this, we call it alternating current, or AC.

> QUESTION: What is the name for a current that reverses direction on a regular basis? (T5A09)
>
> ANSWER: **Alternating current**

One of the most important parameters of an alternating current is its frequency. The frequency of an alternating current is the number of cycles per second, where a cycle is the time required for an alternating current to go from 0 V to its most positive value, then negative to its most negative value, then positive to 0 V again.

> QUESTION: What term describes the number of times per second that an alternating current makes a complete cycle? (T5A12)

ANSWER: **Frequency**

QUESTION: What is the unit of frequency? (T5C05)

ANSWER: **Hertz**

1 Hz is equal to one cycle per second. An alternating current reverses polarity twice per cycle, so the frequency of the alternating current available from a wall socket is 60 Hz.

Conductors are materials that conduct electrical current well or, in other words, have a low resistance. We use copper wires to connect a power supply to a radio because copper wires are good conductors.

QUESTION: Which of the following is a good electrical conductor? (T5A07)

ANSWER: **Copper**

Silver is actually a better conductor than copper, but copper is a lot less expensive than silver. Often, you will see gold used as a conductor. Although gold is not as good a conductor as either copper or silver, it doesn't corrode like copper or silver. That makes it a good choice for switch or connector contacts.

Many times we need a material that does not conduct current very well. We call these materials insulators, and insulators have a high resistance. Plastics and glass are commonly used insulators.

QUESTION: Which of the following is a good electrical insulator? (T5A08)

ANSWER: **Glass**

Ohm's Law: formulas and usage

Hams obey Ohm's Law!

Ohm's Law is the relationship between voltage, current, and resistance in an electrical circuit. When you know any two of these values, you can calculate the third.

The most basic equation for Ohm's Law is $E = I \times R$. In other words, when you know the current flowing through a circuit and the resistance of the circuit, you can calculate the voltage across the circuit by multiplying these two values.

> QUESTION: What formula is used to calculate voltage in a circuit? (T5D02)
>
> ANSWER: **Voltage (E) equals current (I) multiplied by resistance (R)**

Using simple algebra, you can derive the other two forms of this equation. These two equations let you calculate the resistance in a circuit if you know the voltage and current or the current in a circuit if you know the voltage and resistance.

> QUESTION: What formula is used to calculate resistance in a circuit? (T5D03)
>
> ANSWER: **Resistance (R) equals voltage (E) divided by current (I)**

We can also write this formula as $R = E \div I$.

> QUESTION: What formula is used to calculate current in a circuit? (T5D01)
>
> ANSWER: **Current (I) equals voltage (E) divided by resistance (R)**

This formula is written $I = E \div R$.

Now, let's look at some examples of how to apply Ohm's Law.

> QUESTION: What is the resistance of a circuit in which a current of 3 amperes flows through a resistor connected to 90 volts? (T5D04)
>
> ANSWER: **30 ohms**

Here's how to calculate this answer: $R = E \div I = 90\ V \div 3\ A = 30\ \Omega$

> QUESTION: What is the resistance in a circuit for which the applied voltage is 12 volts and the current flow is 1.5 amperes? (T5D05)
>
> ANSWER: **8 ohms**

$R = E \div I = 12\ V \div 1.5\ A = 8\ \Omega$

> QUESTION: What is the resistance of a circuit that draws 4 amperes from a 12-volt source? (T5D06)
>
> ANSWER: **3 ohms**

$R = E \div I = 12\ V \div 4\ A = 3\ \Omega$. Now, let's look at another form of the Ohm's Law equation, $I = E \div R$ to calculate the current in a circuit.

> QUESTION: What is the current in a circuit with an applied voltage of 120 volts and a resistance of 80 ohms? (T5D07)
>
> ANSWER: **1.5 amperes**

$I = E \div R = 120\ V \div 80\ \Omega = 1.5\ A$

> QUESTION: What is the current through a 100-ohm resistor connected across 200 volts? (T5D08)
>
> ANSWER: **2 amperes**

$I = E \div R = 200\ V \div 100\ \Omega = 2\ A$

QUESTION: What is the current through a 24-ohm resistor connected across 240 volts? (T5D09)

ANSWER: **10 amperes**

$I = E \div R = 240\ V \div 24\ \Omega = 10\ A$

Now, let's look at the third form of the Ohm's Law equation, $E = I \times R$ to calculate the voltage across a circuit.

QUESTION: What is the voltage across a 2-ohm resistor if a current of 0.5 amperes flows through it? (T5D10)

ANSWER: **1 volt**

$E = I \times R = 0.5\ A \times 2\ \Omega = 1\ V$

QUESTION: What is the voltage across a 10-ohm resistor if a current of 1 ampere flows through it? (T5D11)

ANSWER: **10 volts**

$E = I \times R = 1\ A \times 10\ \Omega = 10\ V$

QUESTION: What is the voltage across a 10-ohm resistor if a current of 2 amperes flows through it? (T5D12)

ANSWER: **20 volts**

$E = I \times R = 2\ A \times 10\ \Omega = 20\ V$

Series and parallel circuits

Now, let's consider circuits with two resistors instead of just a single resistor. There are two ways in which the two resistors can be connected: in series or in parallel. Figure 2 shows a series circuit.

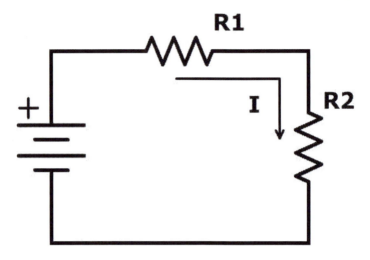

Figure 2. Series circuit.

There is only one path for the current to flow, so the same current flows through both resistors. And, because the voltage across the resistors is equal to I x R, the voltage across each of the resistors will depend on the value of the resistors. If R1 = R2, then the voltage will be the same across both resistors, because the same current flows through both resistors. If R1 does not equal R2, however, the voltages will be different. In either case, the sum of the two voltages will equal the voltage of the voltage source.

> QUESTION: In which type of circuit is current the same through all components? (T5A13)
>
> ANSWER: **Series**
>
> QUESTION: What happens to current at the junction of two components in series? (T5D13)
>
> ANSWER: **It is unchanged**

QUESTION: What is the voltage across each of two components in series with a voltage source? (T5D15)

ANSWER: **It is determined by the type and value of the components**

In a parallel circuit, shown in Figure 3, both resistors are connected directly to the voltage source.

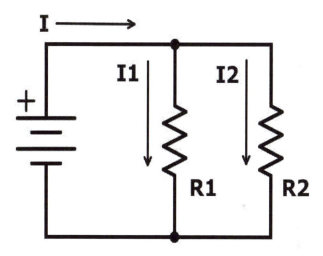

Figure 3. Parallel circuit.

Because both components are connected directly to the voltage source, the voltage across them will be the same. This voltage will cause currents to flow in each of the resistors. I1 = V/R1, and I2 = V/R2. The total current, I, is equal to I1 + I2. If R1 = R2, then the same current flows through both resistors. If the resistors have different values, then I1 will be different from I2.

QUESTION: In which type of circuit is voltage the same across all components? (T5A14)

ANSWER: **Parallel**

QUESTION: What is the voltage across each of two components in parallel with a voltage source? (T5D16)

ANSWER: **The same voltage as the source**

QUESTION: What happens to current at the junction of two components in parallel? (T5D14)

ANSWER: **It divides between them dependent on the value of the components**

DC power

Power is the rate at which electrical energy is generated or consumed. Power is measured in watts. We use the letter P to stand for power and the letter W to stand for watts.

QUESTION: Which term describes the rate at which electrical energy is used? (T5A10)

ANSWER: **Power**

QUESTION: Electrical power is measured in which of the following units? (T5A02)

ANSWER: **Watts**

To calculate power, we multiply the voltage across a circuit by the current flowing through the circuit. We write this equation $P = E \times I$.

QUESTION: What is the formula used to calculate electrical power in a DC circuit? (T5C08)

ANSWER: **Power (P) equals voltage (E) multiplied by current (I)**

Here are some examples:

QUESTION: How much power is being used in a circuit when the applied voltage is 13.8 volts DC and the current is 10 amperes? (T5C09)

ANSWER: **138 watts**

The calculation for this question is $P = E \times I = 13.8 \text{ V} \times 10 \text{ A} = 138 \text{ W}$.

QUESTION: How much power is being used in a circuit when the applied voltage is 12 volts DC and the current is 2.5 amperes? (T5C10)

ANSWER: **30 watts**

The calculation for this question is P = E × I = 12 V × 2.5 A = 30 W.

Just as with Ohm's Law, you can use algebra to come up with other forms of this equation to calculate the voltage if you know the power and the current, or to calculate the current if you know the power and the voltage. The formula to calculate the current, if you know the power and the voltage, is I = P ÷ E.

> QUESTION: How many amperes are flowing in a circuit when the applied voltage is 12 volts DC and the load is 120 watts? (T5C11)
>
> ANSWER: **10 amperes**

In this question, "load" means the power the circuit is consuming. The calculation for this question is I = P ÷ E = 120 W ÷ 12 V = 10 A.

E = VOLTAGE
I = CURRENT
R = RESISTANCE

Math for electronics and conversion of electrical units

When dealing with electrical parameters such as voltage, resistance, current, and power, we use a set of prefixes to denote various orders of magnitude:

- milli- is the prefix used to denote 1 one-thousandth of a quantity. A milliampere, for example, is 1 one-thousandth of an ampere, or 0.001 A. Often, the letter m is used instead of the prefix milli-. 1 milliampere is, therefore, 1 mA.
- micro- is the prefix used to denote 1 one-millionth of a quantity. A microvolt, for example, is 1 one-millionth of a volt, or 0.000001 V. Often, you will see the Greek letter mu, or μ, used to denote the prefix micro-. 1 microvolt is, therefore, 1 μV.
- pico- is the prefix used to denote 1 one-trillionth of a quantity. A picovolt is 1 one-trillionth of a volt, or 0.000001 μV.
- kilo- is the prefix used to denote 1 thousand of a quantity. A kilovolt, for example, is 1000 volts. Often, the letter k is used instead of the prefix kilo-. 1 kilovolt is, therefore, 1 kV.
- mega- is the prefix used to denote 1 million of a quantity. A megahertz, for example, is 1 million Hertz. Often, the letter M is used instead of the prefix mega-. 1 megahertz is, therefore, 1 MHz.
- giga - is the prefix used to denote one billion of a quantity. One gigahertz, or 1 GHz, for example is 1 billion Hertz.

Prefix	Abbreviation	Numerical	Exponential
giga-	G	1,000,000,000	10^9
mega-	M	1,000,000	10^6
kilo-	k	1,000	10^3
----	----	1	10^0
milli-	m	0.001	10^{-3}
micro-	μ,u	0.000001	10^{-6}
nano-	n	0.000000001	10^{-9}
pico-	p	0.000000000001	10^{-12}

Here are some examples:

QUESTION: How many milliamperes is 1.5 amperes? (T5B01)

ANSWER: **1500 milliamperes**

To convert amperes to milliamperes, you multiply by 1,000.

QUESTION: What is another way to specify a radio signal frequency of 1,500,000 hertz? (T5B02)

ANSWER: **1500 kHz**

To convert from hertz (Hz) to kHz, you divide by 1,000.

QUESTION: How many volts are equal to one kilovolt? (T5B03)

ANSWER: **One thousand volts**

QUESTION: How many volts are equal to one microvolt?

(T5B04)

ANSWER: **One one-millionth of a volt**

To convert from kilovolts to volts, you multiply by 1,000. To convert from volts to microvolts, you divide by one million.

QUESTION: Which of the following is equal to 500 milliwatts? (T5B05)

ANSWER: **0.5 watts**

To convert from milliwatts to watts, you divide by 1,000. 500 ÷ 1000 = ½ or 0.5.

QUESTION: How many microfarads are equal to 1,000,000 picofarads? (T5B08)

ANSWER: **1 microfarad**

The farad is the unit of capacitance. There are 1 million picofarads in a microfarad.

QUESTION: If an ammeter calibrated in amperes is used to measure a 3000-milliampere current, what reading would it show? (T5B06)

ANSWER: **3 amperes**

There are a thousand milliamperes in an ampere, so to convert from milliamperes to amperes, you divide by 1,000.

QUESTION: What is the proper abbreviation for megahertz? (T5C14)

ANSWER: **MHz**

QUESTION: If a frequency display calibrated in megahertz shows a reading of 3.525 MHz, what would it show if it were calibrated in kilohertz? (T5B07)

ANSWER: **3525 kHz**

QUESTION: Which of the following frequencies is equal to 28,400 kHz? (T5B12)

ANSWER: **28.400 MHz**

QUESTION: If a frequency display shows a reading of 2425 MHz, what frequency is that in GHz? (T5B13)

ANSWER: **2.425 GHz**

To convert from MHz to kHz, you multiply by 1,000. To convert from kHz to MHz, or to convert from MHz to GHz, you divide by 1,000.

Decibels

When dealing with ratios—especially power ratios—we often use decibels (dB). The reason for this is that the decibel scale is a logarithmic scale, meaning that we can talk about large ratios with relatively small numbers. When the value is positive, it means that there is a power increase. When the value is negative, it means that there is a power decrease.

At this point, you don't need to know the formula used to calculate the ratio in dB, but you need to know the ratios represented by the values 3 dB, 6 dB, and 10 dB.

> QUESTION: What is the approximate amount of change, measured in decibels (dB), of a power increase from 5 watts to 10 watts? (T5B09)
>
> ANSWER: **3 dB**

3 dB corresponds to a ratio of 2 to 1, and because going from 5 watts to 10 watts doubles the power, we can also say that there is a gain of 3 dB.

> QUESTION: What is the approximate amount of change, measured in decibels (dB), of a power decrease from 12 watts to 3 watts? (T5B10)
>
> ANSWER: **–6 dB**

6 dB corresponds to a ratio of 4 to 1, and a decrease in power from 12 watts to 3 watts is a ratio of 4 to 1. Because this is a power decrease, the value in dB is negative.

> QUESTION: What is the amount of change, measured in decibels (dB), of a power increase from 20 watts to 200 watts? (T5B11)
>
> ANSWER: **10 dB**

Increasing the power from 20 watts to 200 watts is a ratio of 10 to 1, and 10 dB corresponds to a ratio of 10 to 1.

DAN ROMANCHIK, KB6NU

Electronic components and circuit diagrams

Resistors, capacitors and capacitance, inductors and inductance, batteries

Resistors are components that, as the name implies, oppose the flow of current. We use them to control how much current flows in a circuit. The higher the resistance, the lower the current.

Most resistors have a fixed value, specified in ohms, but some are designed to be variable. That is, you can change the resistance of the resistor by turning a shaft or sliding a control back and forth. Variable resistors, also called potentiometers, are often used to allow users to adjust the way a device operates.

> QUESTION: What electrical component opposes the flow of current in a DC circuit? (T6A01)
>
> ANSWER: **Resistor**
>
> QUESTION: What type of component is often used as an adjustable volume control? (T6A02)
>
> ANSWER: **Potentiometer**
>
> QUESTION: What electrical parameter is controlled by a potentiometer? (T6A03)

ANSWER: **Resistance**

Another common electrical component is the capacitor. The most basic type of capacitor consists of two metal plates separated by an insulator, called a dielectric. When you put a DC voltage across a capacitor, an electric current flows into the capacitor until the voltage across the capacitor equals the DC voltage. This puts a positive charge on one plate and a negative charge on the other, thereby creating an electric field between the two plates.

> QUESTION: What type of electrical component consists of two or more conductive surfaces separated by an insulator? (T6A05)
>
> ANSWER: **Capacitor**
>
> QUESTION: What electrical component stores energy in an electric field? (T6A04)
>
> ANSWER: **Capacitor**
>
> QUESTION: What is the ability to store energy in an electric field called? (T5C01)
>
> ANSWER: **Capacitance**
>
> QUESTION: What is the basic unit of capacitance? (T5C02)
>
> ANSWER: The **farad**

The third most common type of electrical component in amateur radio equipment is the inductor. Inductors are usually small coils of wire, and when a current flows through that coil of wire, a magnetic field is set up around the coil.

> QUESTION: What electrical component usually is constructed as a coil of wire? (T6A07)

ANSWER: **Inductor**

QUESTION: What type of electrical component stores energy in a magnetic field? (T6A06)

ANSWER: **Inductor**

QUESTION: What is the ability to store energy in a magnetic field called? (T5C03)

ANSWER: **Inductance**

QUESTION: What is the basic unit of inductance? (T5C04)

ANSWER: **The henry**

As amateur radio operators, we often use batteries to power our radio equipment. Some types of batteries are rechargeable, while others are not.

QUESTION: Which of the following battery types is not rechargeable? (T6A11)

ANSWER: **Carbon-zinc**

QUESTION: Which of the following battery types is rechargeable? (T6A10)

ANSWER: **All of these choices are correct**

- Nickel-metal hydride
- Lithium-ion
- Lead-acid gel-cell

There are a couple of random questions about other types of components.
QUESTION: What electrical component is used to connect or disconnect electrical circuits? (T6A08)

ANSWER: **Switch**

QUESTION: What electrical component is used to protect other circuit components from current overloads? (T6A09)

ANSWER: **Fuse**

Semiconductors: basic principles and applications of solid state devices, diodes and transistors

Diodes are the most basic semiconductor component. They have only two electrodes, called the anode and cathode, and conduct current only when it is forward biased. That is to say, diodes only conduct current when the voltage on the anode is positive with respect to the cathode. When the diode is reverse biased, i.e., when the voltage on the anode is negative with respect to the cathode, the diode will not conduct current.

> QUESTION: What electronic component allows current to flow in only one direction? (T6B02)
>
> ANSWER: **Diode**
>
> QUESTION: What are the names of the two electrodes of a diode? (T6B09)
>
> ANSWER: **Anode and cathode**
>
> QUESTION: How is the cathode lead of a semiconductor diode often marked on the package? (T6B06)
>
> ANSWER: **With a stripe**

Light-emitting diodes, or LEDs, are a particular type of diode. When current flows through them, they emit visible light, making them useful as indicators and as part of digital readouts.

> QUESTION: What does the abbreviation LED stand for? (T6B07)
>
> ANSWER: **Light Emitting Diode**
>
> QUESTION: Which of the following is commonly used as a visual indicator? (T6D07)
>
> ANSWER: **LED**

Transistors are semiconductor components designed to control the current flow through them. They have three leads and one of those leads is used as the control pin. A current, in the case of the bipolar junction transistor, or a voltage, in the case of the field effect transistor, on the control pin controls the current flow between the two other pins.

In some circuits, the transistor is used as a switch, and the control signal simply switches the current on and off. In other circuits, the transistor is used as an amplifier. When used as an amplifier, the current through the transistor is proportional to the input current or voltage. The ratio of output current to input current is called the gain of the transistor.

> QUESTION: What class of electronic components uses a voltage or current signal to control current flow? (T6B01)
>
> ANSWER: **Transistors**
>
> QUESTION: Which of these components can be used as an electronic switch or amplifier? (T6B03)
>
> ANSWER: **Transistor**
>
> QUESTION: Which of the following electronic components can amplify signals? (T6B05)
>
> ANSWER: **Transistor**
>
> QUESTION: What is the term that describes a device's ability to amplify a signal? (T6B11)
>
> ANSWER: **Gain**
>
> QUESTION: Which of the following could be the primary gain-producing component in an RF power amplifier? (T6B10)
>
> ANSWER: **Transistor**

Bipolar junction transistors are made up of layers of semiconductor materials that are either P-type, which means that it has a positive net charge, or N-type, which means it has a net negative charge. Each layer has an electrode, making the transistor a device with three leads.

There are two types of bipolar junction transistors: PNP or NPN. A PNP transistor has two P layers, with an N layer sandwiched between them. An NPN transistor has two N layers, with a P layer sandwiched between them.

> QUESTION: Which of the following components can consist of three layers of semiconductor material? (T6B04)
>
> ANSWER: **Transistor**

Another type of transistor often found in amateur radio equipment is the field-effect transistor, or FET. To control the flow of current through the field effect transistor, you use a voltage signal on the control pin. This voltage sets up an electric field inside the transistor, hence the name field effect transistor. FETs, like NPN and PNP transistors have three leads.

> QUESTION: What does the abbreviation FET stand for? (T6B08)
>
> ANSWER: **Field Effect Transistor**

Circuit diagrams, schematic symbols, component functions

When describing circuits on paper, we draw diagrams called schematic diagrams that show the components used in a circuit and how those components are connected together. Using schematic diagrams lets us describe how a circuit works and lets us reproduce a circuit more easily. The circuit components are represented by symbols that readily identify the type of component and its value or part number.

QUESTION: What is the name of an electrical wiring diagram that uses standard component symbols? (T6C01)

ANSWER: **Schematic**

QUESTION: What do the symbols on an electrical schematic represent? (T6C12)

ANSWER: **Electrical components**

QUESTION: Which of the following is accurately represented in electrical schematics? (T6C13)

ANSWER: **The way components are interconnected**

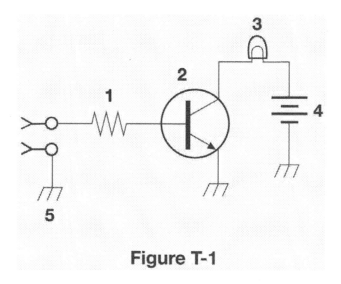

Figure T-1

Figure T-1 is a schematic diagram of a simple transistor circuit. In this circuit, the transistor is used as a switch that turns on a lamp when a positive voltage is applied to the input.

QUESTION: What is component 1 in figure T1? (T6C02)

ANSWER: **Resistor**

Its function is to limit the input current.

QUESTION: What is component 2 in figure T1? (T6C03)

ANSWER: **Transistor**

QUESTION: What is the function of component 2 in Figure T1? (T6D10)

ANSWER: **Control the flow of current**

QUESTION: What is component 3 in figure T1? (T6C04)

ANSWER: **Lamp**

QUESTION: What is component 4 in figure T1? (T6C05)

ANSWER: **Battery**

This battery supplies the current that lights the lamp.

The circuit shown in Figure T2 is a simple power supply. Component 2 is a fuse.

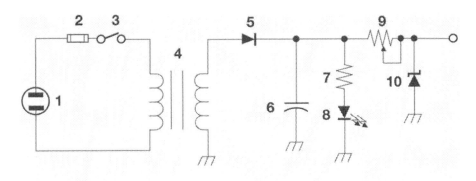

Figure T-2

QUESTION: What type of switch is represented by component 3 in figure T2? (T6D03)

ANSWER: **Single-pole single-throw**

It turns the power supply on and off.

QUESTION: What is component 4 in figure T2? (T6C09)

ANSWER: **Transformer**

QUESTION: What component is commonly used to change 120V AC house current to a lower AC voltage for other uses? (T6D06)

ANSWER: **Transformer**

Component 5 in Figure T2 is a rectifier diode. Rectifier diodes are designed to handle the higher currents found in power supply circuits.

QUESTION: Which of the following devices or circuits changes an alternating current into a varying direct current signal? (T6D01)

ANSWER: **Rectifier**

QUESTION: What is component 6 in figure T2? (T6C06)

ANSWER: **Capacitor**

Sometimes, this is called a filter capacitor, and its function is to filter out remnants of the 60 Hz AC that are part of the varying direct current signal.

QUESTION: What is component 8 in figure T2? (T6C07)

ANSWER: **Light emitting diode**

The LED is a pilot light, serving to alert a user when the power supply is on.

QUESTION: What is component 9 in figure T2? (T6C08)

ANSWER: **Variable resistor**

Its purpose is to limit the output current of the supply.

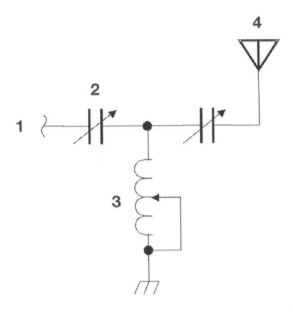

Figure T-3

The circuit shown in Figure T3 is the output circuit of a transmitter.

QUESTION: What is component 3 in figure T3? (T6C10)

ANSWER: **Variable inductor**

There are two variable capacitors in this circuit—component 2 and the unlabeled component.

QUESTION: Which of the following is combined with an inductor to make a tuned circuit? (T6D08)

ANSWER: **Capacitor**

QUESTION: Which of the following is a resonant or tuned circuit? (T6D11)

ANSWER: **An inductor and a capacitor connected in series or parallel to form a filter**

When a capacitor and inductor are connected in series, the circuit has a very low impedance at the resonant frequency. When the capacitor and inductor are connected in parallel, the circuit has a very high impedance at the resonant frequency.

QUESTION: What is component 4 in figure T3? (T6C11)

ANSWER: **Antenna**

Other components

There are many different types of components in modern radio equipment. Below, we will describe the types of components you will need to know about to pass the Technician Class license examination. One of these is a relay.

> QUESTION: What is a relay? (T6D02)
>
> ANSWER: **An electrically-controlled switch**

Meters are devices used to indicate many different values. Meters can indicate the output voltage of a power supply, the output power of a transmitter, and many other values.

> QUESTION: Which of the following displays an electrical quantity as a numeric value? (T6D04)
>
> ANSWER: **Meter**

To make electronic circuits smaller, component manufacturers make devices that have many electronic components on a single piece of silicon. We call these integrated circuits. Integrated circuits, such as microprocessors, may have million of transistors inside them.

> QUESTION: What is the name of a device that combines several semiconductors and other components into one package? (T6D09)
>
> ANSWER: **Integrated circuit**

Integrated circuits may perform either analog or digital functions. One type of analog integrated circuit that is very common is the voltage regulator.

> QUESTION: What type of circuit controls the amount of voltage from a power supply? (T6D05)
>
> ANSWER: **Regulator**

When connecting electronic assemblies together, we often use cables with one or more conductors. Some of those conductors may have a shield around them that is connected to ground.

> QUESTION: Which of the following is a common reason to use shielded wire? (T6D12)
>
> ANSWER: **To prevent coupling of unwanted signals to or from the wire**

DAN ROMANCHIK, KB6NU

Radio wave characteristics

Frequency, wavelength, and the electromagnetic spectrum

Radio waves are what amateur radio is all about. Amateur radio operators generate them and send them off into space. And, on the other side, we capture them and demodulate them.

Radio waves are also called electromagnetic waves because they consist of both an electric wave and a magnetic wave. The two waves are at right angles to one another, and as the wave travels through space, energy gets swapped between the electric and magnetic waves. This is what propels them through space.

> QUESTION: What type of wave carries radio signals between transmitting and receiving stations? (T3A07)
>
> ANSWER: **Electromagnetic**
>
> QUESTION: A radio wave is made up of what type of energy? (T5C07)
>
> ANSWER: **Electromagnetic**
>
> QUESTION: What are the two components of a radio wave? (T3B03)
>
> ANSWER: **Electric and magnetic fields**

One important characteristic of a radio wave is its frequency, or the number of cycles per second. As mentioned earlier, the unit of frequency is the Hertz. We abbreviate Hertz as Hz. One Hz is one cycle per second.

Another important characteristic of a radio wave is the speed at which it travels through space. All electromagnetic waves, no matter the frequency, travel at the speed of light, or 300 million meters per second.

> QUESTION: How fast does a radio wave travel through free space? (T3B04)
>
> ANSWER: **At the speed of light**
>
> QUESTION: What is the approximate velocity of a radio wave as it travels through free space? (T3B11)
>
> ANSWER: **300,000,000 meters per second**
>
> QUESTION: What is the name for the distance a radio wave travels during one complete cycle? (T3B01)
>
> ANSWER: **Wavelength**

Because radio waves travel at the speed of light, or 300,000,000 meters per second, the wavelength is 300,000,000 divided by the frequency. To make this easier to calculate, you can divide both the speed of light and the frequency by one million. That makes the equation:

wavelength (m) = 300/f (MHz)

The converse of this equation is:

f (MHz) = 300/wavelength (m)

As you can see from the equations, the lower the frequency, the longer the wavelength, and vice versa, the higher the frequency, the shorter the wavelength.

> QUESTION: What is the formula for converting frequency to approximate wavelength in meters? (T3B06)

ANSWER: **Wavelength in meters equals 300 divided by frequency in megahertz**

QUESTION: How does the wavelength of a radio wave relate to its frequency? (T3B05)

ANSWER: **The wavelength gets shorter as the frequency increases**

In amateur radio, we sometimes use the frequency and sometimes the wavelength when talking about a radio signal. We use wavelength, for example, when we refer to the amateur radio bands.

QUESTION: What property of radio waves is often used to identify the different frequency bands? (T3B07)

ANSWER: **The approximate wavelength**

The 2 m amateur radio band, for example, spans 144 MHz to 148 MHz. A radio wave with a frequency of 148 MHz, would have a wavelength of 2.03 meters.

For convenience, we split the entire range of radio frequencies into sub-ranges, including high frequency (HF), very high frequency (VHF), and ultra high frequency (UHF).

QUESTION: What frequency range is referred to as HF? (T3B10)

ANSWER: **3 to 30 MHz**

QUESTION: What are the frequency limits of the VHF spectrum? (T3B08)

ANSWER: **30 to 300 MHz**

QUESTION: What are the frequency limits of the UHF spectrum? (T3B09)

ANSWER: **300 to 3000 MHz**

A radio signal of any frequency is called a radio frequency, or RF, signal.

QUESTION: What does the abbreviation "RF" refer to? (T5C06)

ANSWER: **Radio frequency signals of all types**

Properties of radio waves and propagation modes

As amateur radio operators, we should always try to use the right frequency and the right mode when communicating. To do this, we need to know how radio signals travel from one point to another and what effect frequency, our antennas and even our location have on signal propagation.

Communications at VHF and UHF frequencies are generally "line-of-sight" communications. This means they normally travel in a straight line from the transmitter to the receiver. For this reason, they are normally used for local communications.

> QUESTION: Why are direct (not via a repeater) UHF signals rarely heard from stations outside your local coverage area? (T3C01)
>
> ANSWER: **UHF signals are usually not reflected by the ionosphere**

We'll talk more about the ionosphere below.

Because VHF and UHF signals are line-of-sight, at some distance, the signals will be blocked by the curvature of the Earth. The maximum distance for line-of-sight communications is called the radio horizon. The radio horizon extends somewhat farther than the visual horizon.

> QUESTION: Why do VHF and UHF radio signals usually travel somewhat farther than the visual line of sight distance between two stations? (T3C11)
>
> ANSWER: **The Earth seems less curved to radio waves than to light**

One problem often encountered when using VHF and UHF frequencies is multi-path distortion. Multi-path distortion occurs when your signals arrive at a receiving station via two or more paths. Since the signal paths may be different lengths, the signals may arrive out of phase and cancel one another.

> QUESTION: What should you do if another operator reports that your station's 2 meter signals were strong just a moment ago, but now they are weak or distorted? (T3A01)
>
> ANSWER: **Try moving a few feet or changing the direction of your antenna, if possible, as reflections may be causing multi-path distortion**

Moving a few feet might eliminate or reduce the effects of the random reflections that are causing multi-path distortion.

Multi-path distortion affects both voice and digital transmissions.

> QUESTION: What may occur if data signals arrive via multiple paths? (T3A10)
>
> ANSWER: **Error rates are likely to increase**

Knowing how VHF and UHF signals propagate can help you communicate even in adverse conditions. When trying to use a repeater, for example, you may find yourself in a place where a direct path to the repeater is not possible. If you find yourself in this situation, you could try using a directional antenna and bounce your signal off buildings or other obstructions.

> QUESTION: When using a directional antenna, how might your station be able to access a distant repeater if buildings or obstructions are blocking the direct line-of-sight path? (T3A05)
>
> ANSWER: **Try to find a path that reflects signals to the repeater**

Another phenomenon you might use when a direct path to a repeater is not possible is "knife-edge" diffraction. You might be able to use this phenomenon to get your signal around a building in an urban setting.

QUESTION: Which of the following effects might cause radio signals to be heard despite obstructions between the transmitting and receiving stations? (T3C05)

ANSWER: **Knife-edge diffraction**

Mobile operation has its own unique challenges as your transmitter location is constantly changing. This means that the signal at the receiving station constantly changes as well.

QUESTION: What term is commonly used to describe the rapid fluttering sound sometimes heard from mobile stations that are moving while transmitting? (T3A06)

ANSWER: **Picket fencing**

Another condition that could impede the transmission of VHF and UHF signals is vegetation. So, keep your antennas out of trees or above trees.

QUESTION: Why might the range of VHF and UHF signals be greater in the winter? (T3A02)

ANSWER: **Less absorption by vegetation**

Antenna polarization is important at VHF and UHF frequencies. How you mount an antenna directly affects its polarization

QUESTION: What property of a radio wave is used to describe its polarization? (T3B02)

ANSWER: **The orientation of the electric field**

QUESTION: What can happen if the antennas at opposite ends of a VHF or UHF line of sight radio link are not using the same polarization? (T3A04)

ANSWER: **Signals could be significantly weaker**

When using a repeater, vertical polarization is most often used. So, when using a handheld transceiver, make sure to hold it so that your antenna is vertically oriented.

Different activities use different antenna polarizations.

> QUESTION: What antenna polarization is normally used for long-distance weak-signal CW and SSB contacts using the VHF and UHF bands? (T3A03)
>
> ANSWER: **Horizontal**

The reason for this is that weak signal operators are often using what are called beam antennas, and it's much easier to mount and operate beam antennas horizontally than it is to mount them vertically.

Even though VHF communications are most often line-of-sight, there are times when it's possible to communicate over long distances. Sometimes, VHF signals will bounce off the E layer of the ionosphere. This phenomenon is called "sporadic E" because it happens only sporadically.

> QUESTION: Which of the following propagation types is most commonly associated with occasional strong over-the-horizon signals on the 10, 6, and 2 meter bands? (T3C04)
>
> ANSWER: **Sporadic E**

Other interesting propagation phenomena at VHF frequencies include auroral reflection, meteor scatter, tropospheric scatter, and tropospheric ducting. Bouncing signals off the earth's aurora is very interesting.

> QUESTION: What is a characteristic of VHF signals received via auroral reflection? (T3C03)
>
> ANSWER: **The signals exhibit rapid fluctuations of strength and often sound distorted**

Some hams also bounce signals off meteor showers. This propagation mode is called meteor scatter.

> QUESTION: What band is best suited to communicating via meteor scatter? (T3C07)
>
> ANSWER: **6 meter band**

One question that I get from people not knowledgeable about amateur radio is whether or not the weather affects radio wave propagation. The short answer is no, but the exception to the rule is tropospheric ducting or tropospheric scatter. The troposphere is the lowest region of the atmosphere, extending from the earth's surface to a height of about 6–10 km.

> QUESTION: What mode is responsible for allowing over-the-horizon VHF and UHF communications to ranges of approximately 300 miles on a regular basis? (T3C06)
>
> ANSWER: **Tropospheric scatter**
>
> QUESTION: What causes tropospheric ducting? (T3C08)
>
> ANSWER: **Temperature inversions in the atmosphere**

Tropospheric ducting can also propagate VHF signals for many hundreds of miles.

Another exception to the rule occurs at microwave frequencies. Precipitation, including rain, snow, or ice can absorb microwave signals, especially at frequencies above 11 GHz. This phenomenon is called rain fade.

> QUESTION: What weather condition would decrease range at microwave frequencies? (T3A13)
>
> ANSWER: **Precipitation**

At lower frequencies, precipitation has little or no effect.

QUESTION: How might fog and light rain affect radio range on the 10 meter and 6 meter bands? (T3A12)

ANSWER: **Fog and light rain will have little effect on these bands**

HF Propagation

For reliable long-distance communications, amateurs use the HF frequencies. The reason for this is that HF signals bounce off the ionosphere. This phenomenon allows amateur radio operators to contact other amateur radio stations around the world.

The ionosphere is created by solar radiation, which creates a high concentration of ions and free electrons that reflect radio waves. It extends from about 50 to 600 miles above the earth's surface. There are three ionospheric layers—the D, E, and F layers—with the D layer being closest to the Earth, and the F layer being the layer farthest from the surface of the Earth.

> QUESTION: Which part of the atmosphere enables the propagation of radio signals around the world? (T3A11)
>
> ANSWER: **The ionosphere**
>
> QUESTION: Which of the following is an advantage of HF vs VHF and higher frequencies? (T3C02)
>
> ANSWER: **Long distance ionospheric propagation is far more common on HF**

One interesting phenomenon that is related to HF propagation is the sunspot cycle. Generally, the number of sunspots increases and decreases over an 11-year cycle, and HF propagation, especially on the higher frequency HF bands, is best at times when there are many sunspots.

> QUESTION: Which of the following bands may provide long distance communications during the peak of the sunspot cycle? (T3C10)
>
> ANSWER: **6 or 10 meter bands**

Because of the way that the ionosphere changes throughout the day, propagation is best on the higher frequency bands (10m, 15m and 20m) during the day, while propagation is best on the lower frequency bands (160m, 80m, 40m) at night.

QUESTION: What is generally the best time for long-distance 10 meter band propagation via the F layer? (T3C09)

ANSWER: **From dawn to shortly after sunset during periods of high sunspot activity**

A common phenomenon of HF signal propagation is fading.

QUESTION: Which of the following is a likely cause of irregular fading of signals received by ionospheric reflection? (T3A08)

ANSWER: **Random combining of signals arriving via different paths**

This is similar to multi-path distortion of VHF and UHF signals, but in this case, the signals are bouncing off the ionosphere, and because the ionosphere is constantly changing, signals fade in and out.

Antenna polarization is not as important when operating on the HF bands as it is when operating on the VHF/UHF bands.. This is because signals "skip" off the ionosphere and become neither horizontally polarized, nor vertically polarized, but elliptically polarized.

QUESTION: Which of the following results from the fact that skip signals refracted from the ionosphere are elliptically polarized? (T3A09)

ANSWER: **Either vertically or horizontally polarized antennas may be used for transmission or reception**

Antennas and feed lines

Antenna types and antenna polarization

The most common, and perhaps the simplest, antenna is the half-wave dipole antenna. As the name suggests, it measures close to one half wavelength from one end of the antenna to the other. Because half-wave dipole antennas can be quite long—a 40m half-wave dipole antenna, for example, is about 66 feet long—they are normally mounted horizontally between two vertical supports.

> QUESTION: Which of the following describes a simple dipole oriented parallel to the Earth's surface? (T9A03)
>
> ANSWER: **A horizontally polarized antenna**

When putting up a dipole antenna, you should consider the orientation of the antenna because it radiates better in some directions than others.

> QUESTION: In which direction does a half-wave dipole antenna radiate the strongest signal? (T9A10)
>
> ANSWER: **Broadside to the antenna**

So, for example, if you live in Kansas, you might want to orient the antenna so that the wire runs north to south. The antenna will then radiate best east and west, meaning that you'll be able to cover most of the U.S.

The length of a dipole antenna is actually about 5% shorter than the value that you would calculate using the formula wavelength (m) = 300 / frequency (MHz). The reason for this is that there will be some stray capacitance between the wire and the ground and other objects near the antenna. Let's take a look at an example.

> QUESTION: What is the approximate length, in inches, of a half-wavelength 6 meter dipole antenna? (T9A09)
>
> ANSWER: **112**

Here's how to make this calculation. One half wavelength is 3 m. 3 m x 39.4 in/m = 118.2 inches. Since the length of the antenna is about 5% shorter than this calculated value, the approximate length of the antenna will be 118.2 inches x 0.95, or about 112 inches.

Once you have built a dipole antenna, chances are it won't be resonant on the frequency you want. To tune the antenna, you need to determine its resonant frequency and then make it longer or shorter.

> QUESTION: How would you change a dipole antenna to make it resonant on a higher frequency? (T9A05)
>
> ANSWER: **Shorten it**

Conversely, to make it resonant on a lower frequency, you lengthen the antenna. The final length will depend on the height at which you mount the antenna and any trees or buildings that are nearby.

Perhaps the second-most popular type of amateur radio antenna is the quarter-wave vertical antenna. The radiator of a vertical antenna is mounted perpendicular to the Earth. This makes it vertically-polarized, because the electric field will have the same orientation as the antenna's radiator.

Like the half-wave dipole antenna, the length of a quarter-wave vertical antenna will be about 5% shorter than the calculated quarter wavelength.

> QUESTION: What is the approximate length, in inches, of a

quarter-wavelength vertical antenna for 146 MHz? (T9A08)

ANSWER: **19**

The wavelength of a 146 MHz radio wave is approximately 2 m. One quarter of a wavelength is therefore 0.5 m. 0.5 m x 39.4 in/m = 19.7 in. 19.7 in x 0.95 ≈ 19 in.

Because HF antennas can be very long, many amateurs use a technique called "loading" to shorten them. You can use either inductors or capacitors to load an antenna, but the most common way is to use an inductor. In either case, loading an antenna makes it seem electrically longer to a signal at the antenna feed point.

QUESTION: Which of the following describes a type of antenna loading? (T9A02)

ANSWER: **Inserting an inductor in the radiating portion of the antenna to make it electrically longer**

While this technique does allow you to shorten an antenna, the shortened antenna will not be as efficient as a full-size antenna.

Many amateurs use directional, or beam, antennas. There are many different types of beam antennas, including the Yagi, the quad, and the dish antenna. They focus the power that is applied to them in a particular direction, and are therefore said to have "gain" in that direction. What this means is that they will output more power in a particular direction than say a dipole or vertical antenna.

QUESTION: What is a beam antenna? (T9A01)

ANSWER: **An antenna that concentrates signals in one direction**

QUESTION: What type of antennas are the quad, Yagi, and dish? (T9A06)

ANSWER: **Directional antennas**

QUESTION: What is the gain of an antenna? (T9A11)

ANSWER: **The increase in signal strength in a specified direction compared to a reference antenna**

Most handheld VHF and UHF transceivers come with what's called a "rubber duck" antenna. Rubber duck antennas use inductive loading to make them shorter than a full-sized antenna. Loading them like this makes them less efficient than a full-sized quarter-wavelength vertical antenna.

QUESTION: What is a disadvantage of the "rubber duck" antenna supplied with most handheld radio transceivers when compared to a full-sized quarter-wave antenna? (T9A04)

ANSWER: **It does not transmit or receive as effectively**

QUESTION: What is a disadvantage of using a handheld VHF transceiver, with its integral antenna, inside a vehicle? (T9A07)

ANSWER: **Signals might not propagate well due to the shielding effect of the vehicle**

Think about it. If the antenna is inside the vehicle, the metal frame will prevent some of your signal from getting outside the vehicle and block some of the signal from a repeater or other station from reaching the antenna.

A better option is to use an externally-mounted antenna. A popular choice for externally-mounted mobile antennas is the 5/8-wavelength vertical antenna.

QUESTION: What is an advantage of using a properly mounted 5/8 wavelength antenna for VHF or UHF mobile service? (T9A12)

ANSWER: **It has a lower radiation angle and more gain than a 1/4 wavelength antenna**

Feed lines and connectors

Feed lines connect radios to antennas. There are many different types of feed lines, including coaxial cable, ladder line, twin lead, and open-wire feed line, but coaxial cable is the most common type.

> QUESTION: Why is coaxial cable the most common feed line selected for amateur radio antenna systems? (T9B03)
>
> ANSWER: **It is easy to use and requires few special installation considerations**

Perhaps the most important consideration when choosing a feed line is to match the impedance of the feed line to the output impedance of the transmitter and the input impedance of the antenna.

> QUESTION: What is impedance? (T5C12)
>
> ANSWER: **A measure of the opposition to AC current flow in a circuit**

> QUESTION: What is a unit of impedance? (T5C13)
>
> ANSWER: **Ohms**

Most amateur radio transmitters have an output impedance of 50 ohms, and most antennas have an input impedance close to 50 ohms. Because this is the case, most coaxial cable used in amateur radio stations has an impedance of 50 ohms.

> QUESTION: What is the impedance of most coaxial cables used in amateur radio installations? (T9B02)
>
> ANSWER: **50 ohms**

RG-58 and RG-8 are two types of coaxial cable often used in amateur radio stations. Both have an impedance of 50 ohms.

Coaxial cable does, however, have some disadvantages. One of them is that it may be lossy at high frequencies.

QUESTION: In general, what happens as the frequency of a signal passing through coaxial cable is increased? (T9B05)

ANSWER: **The loss increases**

QUESTION: What is the electrical difference between RG-58 and RG-8 coaxial cable? (T9B10)

ANSWER: **RG-8 cable has less loss at a given frequency**

In general, the smaller the diameter of the coaxial cable, the higher the losses will be at higher frequencies. And, the longer the feed line, the higher the losses. So, for example, using RG-58 coax as the feed line for an 80 m dipole might be just fine, but you don't want to use 50 feet of RG-58 to connect your 440 MHz FM transceiver to an antenna on your roof or on a tower.

Even RG-8 coaxial cable is not really the best choice for use at VHF and UHF frequencies. Most repeaters, for example, use air-insulated hard line coaxial cable for the feed line.

QUESTION: Which of the following types of feed line has the lowest loss at VHF and UHF? (T9B11)

ANSWER: **Air-insulated hard line**

Another disadvantage of coaxial cable is that it is more sensitive to moisture than other types of feed lines.

QUESTION: Which of the following is the most common cause for failure of coaxial cables? (T7C09)

ANSWER: **Moisture contamination**

One way that moisture enters a cable is via cracks in the cable's outer jacket.

QUESTION: Why should the outer jacket of coaxial cable be resistant to ultraviolet light? (T7C10)

ANSWER: **Ultraviolet light can damage the jacket and allow water to enter the cable**

Air-insulated hard line is also susceptible to moisture problems.

QUESTION: What is a disadvantage of air core coaxial cable when compared to foam or solid dielectric types? (T7C11)

ANSWER: **It requires special techniques to prevent water absorption**

PL-259 connectors are the most common type of connectors used on coaxial cables in amateur radio stations.

QUESTION: Which of the following is true of PL-259 type coax connectors? (T9B07)

ANSWER: **They are commonly used at HF frequencies**

One problem with PL-259 connectors is that they are not the most suitable connector when operating at higher frequencies. Type N connectors are a much better choice for UHF frequencies.

QUESTION: Which of the following connectors is most suitable for frequencies above 400 MHz? (T9B06)

ANSWER: **A Type N connector**

Another problem with PL-259 connectors is that they are not waterproof. If installed outdoors, water can get into the coaxial cable if it is not sealed in some way. This can lead to increased loss and eventual failure.

QUESTION: Why should coax connectors exposed to the weather be sealed against water intrusion? (T9B08)

ANSWER: **To prevent an increase in feed line loss**

Coaxial cable connectors can be a real pain to install properly and are a frequent cause of failure. When installing a feed line, make sure that your coaxial connectors are soldered properly and that they make solid connections with one another.

QUESTION: What can cause erratic changes in SWR readings? (T9B09)

ANSWER: **A loose connection in an antenna or a feed line**

Standing wave ratio and antenna measurements

Standing wave ratio, or SWR, is a term you'll often hear when talking about antennas and feed lines. It is a measure of how well-matched a feed line is to an antenna. When we say that an antenna is matched to a transmission line, we mean that the impedance of the transmission line is equal to the impedance of the antenna.

> QUESTION: What, in general terms, is standing wave ratio (SWR)? (T7C03)
>
> ANSWER: **A measure of how well a load is matched to a transmission line**

In this context, the "load" is the antenna.

The higher the SWR, the greater the mismatch between the antenna and the transmission line.

> QUESTION: What reading on an SWR meter indicates a perfect impedance match between the antenna and the feed line? (T7C04)
>
> ANSWER: **1 to 1**
>
> QUESTION: What does an SWR reading of 4:1 indicate? (T7C06)
>
> ANSWER: **Impedance mismatch**

When the feed line impedance matches the antenna input impedance, the SWR will be low. The reason this is a good thing is that when the feed line impedance matches the antenna input impedance, the most power is transferred to the antenna and radiated.

> QUESTION: Why is it important to have low SWR when using coaxial cable feed line? (T9B01)
>
> ANSWER: **To reduce signal loss**

The bigger the mismatch is between the feed line and the load, the higher the SWR will be. The higher the SWR, the more power you will lose in the feed line.

> QUESTION: What happens to power lost in a feed line? (T7C07)
>
> ANSWER: **It is converted into heat**

Power converted into heat is not radiated by the antenna, meaning your radiated signal will be weaker.

You can measure the SWR of your antenna system with an SWR meter. You usually connect the SWR meter between the transmitter and antenna, near the output of your transmitter because it is important to have a low SWR at that point.

> QUESTION: What is the proper location for an external SWR meter? (T4A05)
>
> ANSWER: **In series with the feed line, between the transmitter and antenna**

Most amateur radio transceivers today will reduce output power should the SWR of the antenna system get too high. The reason for this is to protect the radio from being damaged by reflected power.

> QUESTION: Why do most solid-state amateur radio transmitters reduce output power as SWR increases? (T7C05)
>
> ANSWER: **To protect the output amplifier transistors**

An SWR meter is not the only way to measure SWR.

> QUESTION: What instrument other than an SWR meter could you use to determine if a feed line and antenna are properly matched? (T7C08)

ANSWER: **Directional wattmeter**

When using a directional wattmeter, you first measure the forward power, then the reflected power, and using those two values, calculate the SWR.

Another test instrument that you can use to measure the SWR of an antenna system is the antenna analyzer.

> QUESTION: Which of the following instruments can be used to determine if an antenna is resonant at the desired operating frequency? (T7C02)
>
> ANSWER: **An antenna analyzer**

Some antenna analyzers will also let you measure capacitive reactance and inductive reactance, and also be used as an RF signal generator. These capabilities may be useful in applications other than antenna analysis.

If an antenna's impedance is not 50 ohms, the impedance at the transmitter end of a feed line will not be 50 ohms. Don't worry, though. You can use a device called an antenna tuner or antenna coupler to transform the impedance from whatever it happens to be to 50 ohms. We call this process impedance matching.

> QUESTION: What is the major function of an antenna tuner (antenna coupler)? (T9B04)
>
> ANSWER: **It matches the antenna system impedance to the transceiver's output impedance**

In addition to instruments that make antenna measurements, it's helpful to have an instrument that can simulate an antenna. That's the purpose of a dummy load. Basically, a dummy load is just a big 50 Ω resistor that provides a known impedance to the transmitter and converts the transmitter output power into heat so that it does not get radiated. If a transmitter operates normally when a dummy load is connected to it, you can be reasonably sure that your transmitter is good.

QUESTION: What is the primary purpose of a dummy load? (T7C01)

ANSWER: **To prevent transmitting signals over the air when making tests**

QUESTION: What does a dummy load consist of? (T7C12)

ANSWER: **A non-inductive resistor and a heat sink**

Amateur radio signals

Modulation modes and signal bandwidth

Modulation is the process of adding information to a radio wave. For example, to send a voice signal, you modulate the radio wave with your voice. There are several different types of modulation, but when you get your Technician license, chances are that frequency modulation, or FM, is the type of modulation that you'll use first. When you frequency modulate a radio, the audio actually changes the frequency of the radio wave a little in proportion to the volume of the audio signal. An FM receiver detects this frequency change and recreates the audio.

> QUESTION: Which type of modulation is most commonly used for VHF and UHF voice repeaters? (T8A04)
>
> ANSWER: **FM**
>
> QUESTION: What type of modulation is most commonly used for VHF packet radio transmissions? (T8A02)
>
> ANSWER: **FM**

Amplitude modulation, or AM, is another type of modulation. To amplitude modulate a signal, you vary the amplitude of the signal in proportion to the audio level. Single-sideband, or SSB, is a form of amplitude modulation.

> QUESTION: Which of the following is a form of amplitude modulation? (T8A01)
>
> ANSWER: **Single sideband**
>
> QUESTION: Which type of voice mode is most often used for long-distance (weak signal) contacts on the VHF and UHF bands? (T8A03)
>
> ANSWER: **SSB**
>
> QUESTION: Where may SSB phone be used in amateur bands above 50 MHz? (T2B13)
>
> ANSWER: **In at least some portion of all these bands**

A single-sideband signal may be upper-sideband (USB) or lower-sideband (LSB).

> QUESTION: Which sideband is normally used for 10 meter HF, VHF, and UHF single-sideband communications? (T8A06)
>
> ANSWER: **Upper sideband**

The audio of an FM signal sounds better than an AM signal, but it occupies a wider bandwidth than an AM or single sideband signal. This means that you can have fewer FM signals in a given frequency range than SSB signals.

> QUESTION: What is an advantage of single sideband (SSB) over FM for voice transmissions? (T8A07)
>
> ANSWER: **SSB signals have narrower bandwidth**
>
> QUESTION: What is the approximate bandwidth of a single sideband (SSB) voice signal? (T8A08)
>
> ANSWER: **3 kHz**

QUESTION: What is the approximate bandwidth of a VHF repeater FM phone signal? (T8A09)

ANSWER: **Between 10 and 15 kHz**

Some modes have very wide bandwidths, such as analog, fast-scan TV.

QUESTION: What is the typical bandwidth of analog fast-scan TV transmissions on the 70 centimeter band? (T8A10)

ANSWER: **About 6 MHz**

QUESTION: What type of transmission is indicated by the term "NTSC?" (T8D04)

ANSWER: **An analog fast scan color TV signal**

NTSC stands for National Television Standards Committee.

Morse Code, or CW, signals have a narrower bandwidth than either AM or FM.

QUESTION: Which of the following types of emission has the narrowest bandwidth? (T8A05)

ANSWER: **CW**

QUESTION: What is the approximate maximum bandwidth required to transmit a CW signal? (T8A11)

ANSWER: **150 Hz**

Technicians have CW privileges on the 80m, 40m, 15m, and 10m bands, as well as all the VHF, UHF, and microwave bands.

QUESTION: What code is used when sending CW in the amateur bands? (T8D09)

ANSWER: **International Morse**

QUESTION: What is an electronic keyer? (T8D14)

ANSWER: **A device that assists in manual sending of Morse**

code

Digital modes: packet, PSK31

When hams talk about "digital modes," we are talking about modes that send digital data rather than voice or other types of analog signals, such as television. Usually, we connect our transceivers to a computer to modulate and demodulate the digital signals, but some transceivers can do this internally.

> QUESTION: Which of the following is a digital communications mode? (T8D01)
>
> ANSWER: **All of these choices are correct**
>
> - Packet radio
> - IEEE 802.11
> - JT65

Packet radio was one of the first digital modes. It is called packet radio because the data to be sent from station to station are separated into a number of packets which are then sent separately by the transmitting station and received and re-assembled by the receiving station.

> QUESTION: Which of the following may be included in packet transmissions? (T8D08)
>
> ANSWER: **All of these choices are correct**
>
> - A check sum which permits error detection
> - A header that contains the call sign of the station to which the information is being sent
> - Automatic repeat request in case of error

Some amateur radio digital communications systems use protocols which ensure error-free communications. One such system is called an automatic repeat request, or ARQ, transmission system.

> QUESTION: What is an ARQ transmission system? (T8D11)

ANSWER: **A digital scheme whereby the receiving station detects errors and sends a request to the sending station to retransmit the information**

APRS is one service that uses packet radio. The major application of APRS is to send location reports while operating a mobile amateur radio station.

QUESTION: What does the term "APRS" mean? (T8D02)

ANSWER: **Automatic Packet Reporting System**

QUESTION: Which of the following is an application of APRS (Automatic Packet Reporting System)? (T8D05)

ANSWER: **Providing real-time tactical digital communications in conjunction with a map showing the locations of stations**

QUESTION: Which of the following devices is used to provide data to the transmitter when sending automatic position reports from a mobile amateur radio station? (T8D03)

ANSWER: **A Global Positioning System receiver**

A popular digital mode on the HF bands is PSK31. PSK31 signals are modulated by shifting the phase of the tone being sent.

QUESTION: What does the abbreviation "PSK" mean? (T8D06)

ANSWER: **Phase Shift Keying**

The "31" in PSK31 comes from the fact that data is transmitted and received at about 31 baud and that the bandwidth of a PSK31 signal is only about 31 Hz. Fortunately, that is about how fast most people can type.

Digital Mobile Radio, or DMR, is an FM digital communications system that's become quite popular in amateur radio. It allows amateur radio operators to use the spectrum more efficiently by multiplexing two signals in a single 12.5 kHz repeater channel. It also allows amateur radio operators to connect with one another over the internet.

> QUESTION: Which of the following best describes DMR (Digital Mobile Radio)? (T8D07)
>
> ANSWER: **A technique for time-multiplexing two digital voice signals on a single 12.5 kHz repeater channel**

One of the interesting things about operating DMR is the ability to join a "talk group." A talk group is a virtual channel that connects radio users locally over a repeater or state-wide, nationally, or internationally over the internet.

> QUESTION: What is a "talk group" on a DMR digital repeater? (T2B09)
>
> ANSWER: **A way for groups of users to share a channel at different times without being heard by other users on the channel**
>
> QUESTION: How can you join a digital repeater's "talk group"? (T2B07)
>
> ANSWER: **Program your radio with the group's ID or code**

WSJT software is software that provides weak-signal digital communication modes on amateur radio frequencies. To communicate with one of the digital modes that WSJT software provides, you need a single sideband transceiver and a personal computer with a sound card.

> QUESTION: Which of the following operating activities is supported by digital mode software in the WSJT suite? (T8D10)
>
> ANSWER: **All of these choices are correct**
>
> - Moonbounce or Earth-Moon-Earth

- Weak-signal propagation beacons
- Meteor scatter

FT8 is one of the modes supported by WSJT.

QUESTION: What is FT8? (T8D13)

ANSWER: **A digital mode capable of operating in low signal-to-noise conditions that transmits on 15-second intervals**

Mesh networking is another amateur radio activity that's becoming popular. Mesh networks use WiFi frequencies that just happen to be part of the 2.4 GHz, 3.4 GHz, and 5.8 GHz amateur radio bands. Mesh networks allow amateur radio stations to establish digital communications links for supporting high-speed emergency communications and internet connectivity. Broadband-Hamnet and Amateur Radio Emergency Data Network (AREDN) are two projects that use mesh networking to provide digital communications.

QUESTION: Which of the following best describes Broadband-Hamnet(TM), also referred to as a high-speed multi-media network? (T8D12)

ANSWER: **An amateur-radio-based data network using commercial Wi-Fi gear with modified firmware**

Electrical safety

Power circuits and hazards, hazardous voltages, fuses and circuit breakers, grounding, and battery safety

BE SAFE!

When operating or working on amateur radio equipment, it's possible to come into contact with dangerous voltages and currents. People have died working on high-voltage circuits. Because it would be a shame to lose a single person, it's important to know how to be safe when working with electricity.

30 volts is the commonly accepted value for the lowest voltage that can cause a dangerous electric shock, and only 100 mA flowing through the body can cause death. These are not very large values.

> QUESTION: What health hazard is presented by electrical current flowing through the body? (T0A02)
>
> ANSWER: **All of these choices are correct**
> - It may cause injury by heating tissue
> - It may disrupt the electrical functions of cells
> - It may cause involuntary muscle contractions

Start by ensuring that your amateur radio station has three-wire electrical outlets that are properly grounded. When properly wired, three-wire electrical outlets and plugs are safer than two-wire outlets and plugs, and you should use three-wire plugs for all of your amateur radio equipment. The third wire provides an independent, or safety ground.

> QUESTION: In the United States, what is connected to the green wire in a three-wire electrical AC plug? (T0A03)
>
> ANSWER: **Equipment ground**
>
> QUESTION: What is a good way to guard against electrical shock at your station? (T0A06)
>
> ANSWER: **All of these choices are correct**
>
> - Use three-wire cords and plugs for all AC powered equipment
> - Connect all AC powered station equipment to a common safety ground
> - Use a circuit protected by a ground-fault interrupter

Installing fuses in an electrical circuit is the main way to protect a circuit from excessive current flow. Fuses are designed to "blow" or disconnect power if the current flowing through them exceeds a specified value.

> QUESTION: What is the purpose of a fuse in an electrical circuit? (T0A04)
>
> ANSWER: **To interrupt power in case of overload**

When replacing a fuse, always replace the blown fuse with a fuse of the same type and value.

> QUESTION: Why is it unwise to install a 20-ampere fuse in the place of a 5-ampere fuse? (T0A05)
>
> ANSWER: **Excessive current could cause a fire**

If you plan to build your own equipment, be sure to include fuses in your designs.

> QUESTION: What safety equipment should always be included in home-built equipment that is powered from 120V AC power circuits? (T0A08)
>
> ANSWER: **A fuse or circuit breaker in series with the AC hot conductor**

Whenever you're working on equipment, be sure to disconnect it from the power lines, and even then be careful working around a power supply's capacitors.

> QUESTION: What kind of hazard might exist in a power supply when it is turned off and disconnected? (T0A11)
>
> ANSWER: **You might receive an electric shock from the charge stored in large capacitors**

You also need to take precautions when using batteries to power your amateur radio station. Conventional 12-volt storage batteries present several safety hazards.

> QUESTION: Which of the following is a safety hazard of a 12-volt storage battery? (T0A01)
>
> ANSWER: **Shorting the terminals can cause burns, fire, or an explosion**
>
> QUESTION: What can happen if a lead-acid storage battery is charged or discharged too quickly? (T0A10)
>
> ANSWER: **The battery could overheat, give off flammable gas, or explode**

Antenna and tower safety

Antenna safety is also of primary concern. There are two aspects of antenna safety—being safe when installing an antenna and safely operating an antenna. When installing an antenna, make sure you note where the power lines are and stay clear of them.

> QUESTION: Which of the following is an important safety precaution to observe when putting up an antenna tower? (T0B04)
>
> ANSWER: **Look for and stay clear of any overhead electrical wires**
>
> QUESTION: What is the minimum safe distance from a power line to allow when installing an antenna? (T0B06)
>
> ANSWER: **Enough so that if the antenna falls unexpectedly, no part of it can come closer than 10 feet to the power wires**
>
> QUESTION: Why should you avoid attaching an antenna to a utility pole? (T0B09)
>
> ANSWER: **The antenna could contact high-voltage power lines**

You also should position the antenna so that no one can touch it while you are transmitting.

> QUESTION: What could happen if a person accidentally touched your antenna while you were transmitting? (T0C07)
>
> ANSWER: **They might receive a painful RF burn**

There are a number of things you should keep in mind when putting up a tower and installing antennas on them.

QUESTION: When should members of a tower work team wear a hard hat and safety glasses? (T0B01)

ANSWER: **At all times when any work is being done on the tower**

QUESTION: What is a good precaution to observe before climbing an antenna tower? (T0B02)

ANSWER: **Put on a carefully inspected climbing harness (fall arrester) and safety glasses**

QUESTION: Under what circumstances is it safe to climb a tower without a helper or observer? (T0B03)

ANSWER: **Never**

QUESTION: Which of the following is an important safety rule to remember when using a crank-up tower? (T0B07)

ANSWER: **This type of tower must not be climbed unless retracted or mechanical safety locking devices have been installed**

QUESTION: What is the purpose of a safety wire through a turnbuckle used to tension guy lines? (T0B13)

ANSWER: **Prevent loosening of the guy line from vibration**

And, you should use a gin pole designed for use with the tower that you're installing.

QUESTION: What is the purpose of a gin pole? (T0B05)

ANSWER: **To lift tower sections or antennas**

Grounding is very important when installing a tower. The tower is, after all, basically a big lightning rod.

QUESTION: Which of the following establishes grounding requirements for an amateur radio tower or antenna? (T0B11)

ANSWER: **Local electrical codes**

QUESTION: What is considered to be a proper grounding method for a tower? (T0B08)

ANSWER: **Separate eight-foot long ground rods for each tower leg, bonded to the tower and each other**

QUESTION: Which of the following is true when installing grounding conductors used for lightning protection? (T0B10)

ANSWER: **Sharp bends must be avoided**

QUESTION: Which of the following is good practice when installing ground wires on a tower for lightning protection? (T0B12)

ANSWER: **Ensure that connections are short and direct**

QUESTION: What should be done to all external ground rods or earth connections? (T0A09)

ANSWER: **Bond them together with heavy wire or conductive strap**

Lightning can also be conducted down a feed line and into your station. To prevent this, several manufacturers make devices designed to conduct this current to ground before it gets into the station.

QUESTION: Which of these precautions should be taken when installing devices for lightning protection in a coaxial cable feed line? (T0A07)

ANSWER: **Mount all of the protectors on a metal plate that is in turn connected to an external ground rod**

RF hazards and radiation exposure

Over-exposure to radio waves can also be a safety hazard. When using as little as 50 watts, you may be required to perform an RF exposure evaluation for your amateur radio station, even though VHF and UHF radio signals are non-ionizing radiation.

> QUESTION: What type of radiation are VHF and UHF radio signals? (T0C01)
>
> ANSWER: **Non-ionizing radiation**
>
> QUESTION: How does RF radiation differ from ionizing radiation (radioactivity)? (T0C12)
>
> ANSWER: **RF radiation does not have sufficient energy to cause genetic damage**

Even so, small levels of RF energy can be hazardous.

> QUESTION: What is the maximum power level that an amateur radio station may use at VHF frequencies before an RF exposure evaluation is required? (T0C03)
>
> ANSWER: **50 watts PEP at the antenna**

How do you perform an RF exposure evaluation?

> QUESTION: Which of the following is an acceptable method to determine that your station complies with FCC RF exposure regulations? (T0C06)
>
> ANSWER: **All of these choices are correct**
> - By calculation based on FCC OET Bulletin 65
> - By calculation based on computer modeling
> - By measurement of field strength using calibrated equipment

One of the factors to consider when performing an RF exposure evaluation is the duty cycle of your transmissions. A transmission with a lower duty cycle would be less hazardous than a high duty cycle transmission.

> QUESTION: What is the definition of duty cycle during the averaging time for RF exposure? (T0C11)
>
> ANSWER: **The percentage of time that a transmitter is transmitting**
>
> QUESTION: Why is duty cycle one of the factors used to determine safe RF radiation exposure levels? (T0C10)
>
> ANSWER: **It affects the average exposure of people to radiation**

Consider this example.

> QUESTION: If the averaging time for exposure is 6 minutes, how much power density is permitted if the signal is present for 3 minutes and absent for 3 minutes rather than being present for the entire 6 minutes? (T0C13)
>
> ANSWER: **2 times as much**

Because of the way radio waves interact with the body, the exposure limits are different for each amateur radio band.

> QUESTION: Why do exposure limits vary with frequency? (T0C05)
>
> ANSWER: **The human body absorbs more RF energy at some frequencies than at others**
>
> QUESTION: Which of the following frequencies has the lowest value for Maximum Permissible Exposure limit? (T0C02)
>
> ANSWER: **50 MHz**

QUESTION: What factors affect the RF exposure of people near an amateur station antenna? (T0C04)

ANSWER: **All of these choices are correct**

- Frequency and power level of the RF field
- Distance from the antenna to a person
- Radiation pattern of the antenna

So, what should you do if your RF exposure evaluation shows that people are being exposed to excessive RF?

QUESTION: Which of the following actions might amateur operators take to prevent exposure to RF radiation in excess of FCC-supplied limits? (T0C08)

ANSWER: **Relocate antennas**

You could also lower the power or simply transmit less.

You should regularly evaluate your station for RF safety.

QUESTION: How can you make sure your station stays in compliance with RF safety regulations? (T0C09)

ANSWER: **By re-evaluating the station whenever an item of equipment is changed**

DAN ROMANCHIK, KB6NU

Amateur radio practices and station setup

Station setup: connecting microphones, reducing unwanted emissions, power sources, connecting a computer, RF grounding, connecting digital equipment

When setting up an amateur radio station, choosing the radio is the most important consideration, but you must also choose a wide range of accessories, such as power supplies and microphones. In addition, how you connect all these pieces of equipment together is important for your station to operate properly. For example, the cable connecting the power supply to the radio should be made with heavy-gauge wire.

> QUESTION: Why should wiring between the power source and radio be heavy-gauge wire and kept as short as possible? (T4A03)
>
> ANSWER: **To avoid voltage falling below that needed for proper operation**

When choosing a power supply, check the voltage and current ratings of the supply and be sure to choose one capable of supplying a high enough voltage and enough current to power your radio.

> QUESTION: What must be considered to determine the

minimum current capacity needed for a transceiver power supply? (T4A01)

ANSWER: **All of these choices are correct**

- Efficiency of the transmitter at full power output
- Receiver and control circuit power
- Power supply regulation and heat dissipation

A computer has become a common accessory in many amateur radio stations. Amateurs use them to operate digital modes, log contacts, and design circuits and antennas. I even use mine to Twitter while I'm on the air.

QUESTION: How might a computer be used as part of an amateur radio station? (T4A02)

ANSWER: **All of these choices are correct**

- For logging contacts and contact information
- For sending and/or receiving CW
- For generating and decoding digital signals

QUESTION: Which of the following connections might be used between a voice transceiver and a computer for digital operation? (T4A06)

ANSWER: **Receive audio, transmit audio, and push-to-talk (PTT)**

QUESTION: Which computer sound card port is connected to a transceiver's headphone or speaker output for operating digital modes? (T4A04)

ANSWER: **Microphone or line input**

QUESTION: How is a computer's sound card used when conducting digital communications? (T4A07)

ANSWER: **The sound card provides audio to the microphone input and converts received audio to digital form**

Audio and power supply cables in an amateur radio station sometimes pick up stray RF. At minimum, this RF can cause the audio to be noisy. At worst, it can cause a radio or accessory to malfunction.

QUESTION: Which of the following could you use to cure distorted audio caused by RF current on the shield of a microphone cable? (T4A09)

ANSWER: **Ferrite choke**

Good grounding techniques can help you avoid interference problems. When grounding your equipment, you should connect the various pieces of equipment to a single point, keep leads short, and use a heavy conductor to connect to ground.

QUESTION: Which of the following conductors provides the lowest impedance to RF signals? (T4A08)

ANSWER: **Flat strap**

If you plan to install a radio in your car and operate mobile, you have a different set of challenges. One is connecting the radio to the car's power system. Some amateurs connect their radio with a cigarette lighter plug, but this plug is not designed for high currents. For permanent installations, you need to make solid connections at the appropriate points.

QUESTION: Where should the negative return connection of a mobile transceiver's power cable be connected? (T4A11)

ANSWER: **At the battery or engine block ground strap**

The positive connection can also be made at the battery or through an unused position of the vehicle's fuse block.

Another challenge is noise generated by the car itself. The alternator is often the culprit.

> QUESTION: What is the source of a high-pitched whine that varies with engine speed in a mobile transceiver's receive audio? (T4A10)
>
> ANSWER: **The alternator**

Should this be a problem, there are filters that you can install to mitigate the alternator whine. Some HF receivers also have the ability to reduce noise.

> QUESTION: Which of the following could be used to remove power line noise or ignition noise? (T4B12)
>
> ANSWER: **Noise blanker**
>
> QUESTION: Which of the following would reduce ignition interference to a receiver? (T4B05)
>
> ANSWER: **Turn on the noise blanker**

Operating controls: tuning, use of filters, squelch function, AGC, repeater offset, memory channels

To properly operate a transceiver, you need to know how to use the controls. Perhaps the most important transmitter control is microphone gain.

> QUESTION: What may happen if a transmitter is operated with the microphone gain set too high? (T4B01)
>
> ANSWER: **The output signal might become distorted**

You also need to know how to set the operating frequency of your transceiver.

> QUESTION: Which of the following can be used to enter the operating frequency on a modern transceiver? (T4B02)
>
> ANSWER: **The keypad or VFO knob**
>
> QUESTION: What is a way to enable quick access to a favorite frequency on your transceiver? (T4B04)
>
> ANSWER: **Store the frequency in a memory channel**

Transceivers that allow you to store frequencies in memory often have the ability to step through those frequencies, one at a time, stopping when a signal is received. This is called scanning and is a way to monitor many different frequencies automatically.

> QUESTION: Which of the following is a use for the scanning function of an FM transceiver? (T4B13)
>
> ANSWER: **To scan through a range of frequencies to check for activity**

A common receiver control on VHF/UHF transceivers is the squelch control.

> QUESTION: What is the purpose of the squelch control on a transceiver? (T4B03)

ANSWER: **To mute receiver output noise when no signal is being received**

If the squelch control is set too low, the radio will sound noisy. On the other hand, if the squelch control is set too high, then you will not be able to hear low-level signals.

Another common setting on VHF/UHF transceivers is the offset frequency. This is especially important when operating repeaters.

QUESTION: Which is meant by "repeater offset"? (T2A07)

ANSWER: **The difference between the repeater's transmit frequency and its receive frequency**

A common receiver control on HF transceivers is the Receiver Incremental Tuning, or RIT, control. Its purpose is to set the receive frequency slightly off from the transmit frequency.

QUESTION: What does the term "RIT" mean? (T4B07)

ANSWER: **Receiver Incremental Tuning**

QUESTION: Which of the following controls could be used if the voice pitch of a single-sideband signal seems too high or low? (T4B06)

ANSWER: **The receiver RIT or clarifier**

Because HF signals may fade in and out, HF transceivers have a feature called automatic gain control (AGC). AGC eliminates the need for the operator to continually adjust the volume.

QUESTION: What is the function of automatic gain control, or AGC? (T4B11)

ANSWER: **To keep received audio relatively constant**

HF transceivers are often equipped with a variety of different filters. Using the appropriate filter for the mode you are operating can make operating a lot easier.

QUESTION: What is the advantage of having multiple receive bandwidth choices on a multimode transceiver? (T4B08)

ANSWER: **Permits noise or interference reduction by selecting a bandwidth matching the mode**

QUESTION: Which of the following is an appropriate receive filter bandwidth for minimizing noise and interference for SSB reception? (T4B09)

ANSWER: **2400 Hz**

QUESTION: Which of the following is an appropriate receive filter bandwidth for minimizing noise and interference for CW reception? (T4B10)

ANSWER: **500 Hz**

A common transmitter control is push-to-talk, or PTT.

QUESTION: What is meant by PTT? (T7A07)

ANSWER: **The push to talk function that switches between receive and transmit**

Most of the time PTT refers to an actual switch on the microphone that an operator must push to begin transmitting, but it also refers to the name of a signal line on a transceiver's accessory socket that can be used to automatically switch a transceiver into transmit mode.

DAN ROMANCHIK, KB6NU

Station equipment

Receivers, transmitters, transceivers, modulation, transverters, low power and weak signal operation, transmit and receive amplifiers

In the early days of radio, amateur radio operators used separate receivers and transmitter units. Nowadays, however, most use radios called transceivers.

> QUESTION: What is a transceiver? (T7A02)
>
> ANSWER: **A unit combining the functions of a transmitter and a receiver**

Often, HF transceivers are used with devices called transverters that convert the signals from their HF transceiver to the VHF, UHF, and even microwave bands. Transverters take the output of an HF transceiver, normally set to the 10 m (28 MHz) band and output a VHF, UHF, or microwave signal. Conversely, they receive a VHF, UHF, or microwave signal and output a signal in the 10 m band that is demodulated by the HF transceiver.

> QUESTION: What device converts the RF input and output of a transceiver to another band? (T7A06)
>
> ANSWER: **Transverter**

Many, if not most, new amateurs buy a handheld transceiver, called an "HT," as their first transceiver. One disadvantage of using a handheld transceiver is that the maximum output power is generally only 5 W, and because of this, they have limited range. To get around this limitation, you can use an RF amplifier to boost the power.

> QUESTION: What device increases the low-power output from a handheld transceiver? (T7A10)
>
> ANSWER: **An RF power amplifier**
>
> QUESTION: What is the function of the SSB/CW-FM switch on a VHF power amplifier? (T7A09)
>
> ANSWER: **Set the amplifier for proper operation in the selected mode**

When talking about a transceiver's specifications, we still refer to its receiver and transmitter. The two most important specifications for a receiver are sensitivity and selectivity.

> QUESTION: Which term describes the ability of a receiver to detect the presence of a signal? (T7A01)
>
> ANSWER: **Sensitivity**
>
> QUESTION: Which term describes the ability of a receiver to discriminate between multiple signals? (T7A04)
>
> ANSWER: **Selectivity**

To improve the sensitivity of a receiver, you can use an RF preamplifier. An RF preamplifier amplifies signals that you want to receive.

> QUESTION: Where is an RF preamplifier installed? (T7A11)
>
> ANSWER: **Between the antenna and receiver**

Many HF transceivers have some version of a superheterodyne receiver. A superheterodyne receiver converts an incoming radio signal to an intermediate frequency, or IF. The circuit that does this is the mixer.

> QUESTION: Which of the following is used to convert a radio signal from one frequency to another? (T7A03)
>
> ANSWER: **Mixer**

When transmitting or receiving, we want to generate an RF signal with a specific frequency. To do that, we use an oscillator.

> QUESTION: What is the name of a circuit that generates a signal at a specific frequency? (T7A05)
>
> ANSWER: **Oscillator**

To transmit a voice or data signal, we have to combine an audio frequency signal from the microphone with the RF carrier signal generated by the transmitter.

> QUESTION: Which of the following describes combining speech with an RF carrier signal? (T7A08)
>
> ANSWER: **Modulation**

Modulators use a type of mixer circuit to accomplish this process.

Common transmitter and receiver problems: symptoms of overload and overdrive; distortion; causes of interference; interference and consumer electronics; part 15 devices; over-modulation; RF feedback; off frequency signals

Since Murphy's Law—the law that states if anything can go wrong, it will—applies to amateur radio as much as it does to any other pursuit, at some point you will have to deal with problems. These may include overload, distortion, feedback, and interference. Let's first consider interference.

> QUESTION: Which of the following can cause radio frequency interference? (T7B03)
>
> ANSWER: **All of these choices are correct**
>
> - Fundamental overload
> - Harmonics
> - Spurious emissions.

Any of these could cause interference to a TV set or radio or even computer speakers, and you will want to take steps to find and eliminate that interference.

> QUESTION: Which of the following actions should you take if a neighbor tells you that your station's transmissions are interfering with their radio or TV reception? (T7B06)
>
> ANSWER: **Make sure that your station is functioning properly and that it does not cause interference to your own radio or television when it is tuned to the same channel**

While it's not very likely that your amateur radio station will interfere with a neighbor's cable TV service, it can sometimes occur. If you are interfering with a neighbor's cable TV service, first check the connections.

QUESTION: What should be the first step to resolve cable TV interference from your ham radio transmission? (T7B12)

ANSWER: **Be sure all TV coaxial connectors are installed properly**

Your amateur radio station may interfere with a nearby radio receiver if your signal is so strong that the receiver cannot reject the signal even though your signal is not on the frequency to which the receiver is tuned. This is called overload.

QUESTION: What would cause a broadcast AM or FM radio to receive an amateur radio transmission unintentionally? (T7B02)

ANSWER: **The receiver is unable to reject strong signals outside the AM or FM band**

QUESTION: How can overload of a non-amateur radio or TV receiver by an amateur signal be reduced or eliminated? (T7B05)

ANSWER: **Block the amateur signal with a filter at the antenna input of the affected receiver**

The process can work the other way, too. When driving by the antenna of a high-power broadcast station you may notice that your VHF transceiver is picking up the broadcast station signal. This is often the result of overload.

QUESTION: Which of the following can reduce overload to a VHF transceiver from a nearby FM broadcast station? (T7B07)

ANSWER: **Band-reject filter**

The band of frequencies that you want to reject is the band that includes the broadcast station frequency.

Another device that often experiences interference from amateur radio stations is the telephone. The telephone wires act as antenna and the telephone itself demodulates the signal.

> QUESTION: Which of the following is a way to reduce or eliminate interference from an amateur transmitter to a nearby telephone? (T7B04)
>
> ANSWER: **Put a RF filter on the telephone**

Interference works both ways. Your neighbors may have wireless devices, sometimes called "Part 15 devices," that can interfere with your station.

> QUESTION: What is a Part 15 device? (T7B09)
>
> ANSWER: **An unlicensed device that may emit low-powered radio signals on frequencies used by a licensed service**
>
> QUESTION: What should you should if something in a neighbor's home is causing harmful interference to your amateur station? (T7B08)
>
> ANSWER: **All of these choices are correct**
>
> - Work with your neighbor to identify the offending device
> - Politely inform your neighbor about the rules that prohibit the use of devices that cause interference
> - Check your station and make sure it meets the standards of good amateur practice

Perhaps the most common problem that amateur radio operators have is distorted or noisy audio when transmitting. There are many reasons for poor audio.

> QUESTION: What might be a problem if you receive a report that your audio signal through the repeater is distorted

or unintelligible? (T7B10)

ANSWER: **All of these choices are correct**

- Your transmitter is slightly off frequency
- Your batteries are running low
- You are in a bad location

QUESTION: What is a symptom of RF feedback in a transmitter or transceiver? (T7B11)

ANSWER: **Reports of garbled, distorted, or unintelligible voice transmissions**

Sometimes, garbled or distorted audio when operating FM is the result of over-deviation.

QUESTION: What can you do if you are told your FM handheld or mobile transceiver is over-deviating? (T7B01)

ANSWER: **Talk farther away from the microphone**

Basic repair and testing: soldering; using basic test instruments; connecting a voltmeter, ammeter, or ohmmeter

The most common test instrument in an amateur radio shack is the multimeter. Multimeters are called that because they combine the functions of a voltmeter, ohmmeter, and ammeter into a single instrument.

> QUESTION: Which of the following measurements are commonly made using a multimeter? (T7D07)
>
> ANSWER: **Voltage and resistance**

The voltmeter function of the multimeter is used to measure electromotive force, more commonly known as voltage.

> QUESTION: Which instrument would you use to measure electric potential or electromotive force? (T7D01)
>
> ANSWER: **A voltmeter**
>
> QUESTION: What is the correct way to connect a voltmeter to a circuit? (T7D02)
>
> ANSWER: **In parallel with the circuit**
>
> QUESTION: Which of the following precautions should be taken when measuring high voltages with a voltmeter? (T7D12)
>
> ANSWER: **Ensure that the voltmeter and leads are rated for use at the voltages to be measured**

The ohmmeter function of a multimeter is used to measure resistance. The way an ohmmeter measures the resistance of a circuit is by supplying a known current to the circuit, measuring the voltage across the circuit, and then calculating the resistance using Ohm's Law, $R = E/I$.

QUESTION: What instrument is used to measure resistance? (T7D05)

ANSWER: **An ohmmeter**

QUESTION: Which of the following precautions should be taken when measuring circuit resistance with an ohmmeter? (T7D11)

ANSWER: **Ensure that the circuit is not powered**

QUESTION: Which of the following might damage a multimeter? (T7D06)

ANSWER: **Attempting to measure voltage when using the resistance setting**

QUESTION: What is probably happening when an ohmmeter, connected across an unpowered circuit, initially indicates a low resistance and then shows increasing resistance with time? (T7D10)

ANSWER: **The circuit contains a large capacitor**

The ammeter function of a multimeter is used to measure current. You connect an ammeter in series with a circuit so that the current flowing through the circuit also flows through the ammeter.

QUESTION: Which instrument is used to measure electric current? (T7D04)

ANSWER: **An ammeter**

QUESTION: How is a simple ammeter connected to a circuit? (T7D03)

ANSWER: **In series with the circuit**

In addition to knowing how to make electrical measurements, knowing how to solder is an essential skill for amateur radio operators.

QUESTION: Which of the following types of solder is best for radio and electronic use? (T7D08)

ANSWER: **Rosin-core solder**

QUESTION: What is the characteristic appearance of a cold solder joint? (T7D09)

ANSWER: **A grainy or dull surface**

We call a poor solder joint a "cold" solder joint because it's usually the result of not applying enough heat to the joint. When you don't apply enough heat to a solder joint, the solder does not flow smoothly between the metal surfaces to be joined and often does not make a good connection.

Operating procedures

FM Operation

Once they get their licenses, most Technicians purchase a VHF/UHF FM transceiver. This type of radio allows them to use repeaters and participate in public-service events.

QUESTION: What type of amateur station simultaneously retransmits the signal of another amateur station on a different channel or channels? (T1F09)

ANSWER: **Repeater station**

QUESTION: What types of amateur stations can automatically retransmit the signals of other amateur stations? (T1D07)

ANSWER: **Repeater, auxiliary, or space stations**

To use repeaters, you need to know how to set up your radio. Repeaters receive on one frequency and transmit on another. You program your radio so that it receives on the repeater's transmit frequency and transmits on the repeater's receive frequency. The difference between the transmit frequency and receive frequency is called the repeater frequency offset.

QUESTION: Which of the following is a common repeater

frequency offset in the 2 meter band? (T2A01)

ANSWER: **Plus or minus 600 kHz**

QUESTION: What is a common repeater frequency offset in the 70 cm band? (T2A03)

ANSWER: **Plus or minus 5 MHz**

Because repeaters often operate in environments where there is a lot of interference, they are programmed not to operate unless the station they are receiving is also transmitting a sub- audible tone of a specific frequency. If your radio has not been programmed to transmit the proper sub-audible tone when you transmit, the repeater will not repeat your transmission. These tones are sometimes called PL (short for "private line") tones. PL is a Motorola trademark. The generic term for these tones is CTCSS (short for "continuous tone-coded squelch system").

QUESTION: What term describes the use of a sub-audible tone transmitted along with normal voice audio to open the squelch of a receiver? (T2B02)

ANSWER: **CTCSS**

A frequent problem is being able to hear a repeater, but not being able to access it.

QUESTION: Which of the following could be the reason you are unable to access a repeater whose output you can hear? (T2B04)

ANSWER: **All of these choices are correct**

- Improper transceiver offset
- The repeater may require a proper CTCSS tone from your transceiver
- The repeater may require a proper DCS tone from your

transceiver

A Digital Code Squelch, or DCS, tone is similar to a CTCSS tone in that it is sub-audible and opens the squelch of a repeater when a station is trying to access it. They are, however, not very common.

Not using a CTCSS tone, or using the wrong CTCSS tone, is only one of the many problems that you may have when operating through a repeater. Not having a strong enough signal is one of them.

> QUESTION: If a station is not strong enough to keep a repeater's receiver squelch open, which of the following might allow you to receive the station's signal? (T2B03)
>
> ANSWER: **Listen on the repeater input frequency**

One way to listen to the repeater input frequency would be to use the "reverse split" function of your VHF/UHF transceiver, if it has this feature. When enabled, the reverse split feature will cause your transceiver to transmit on the repeater output frequency and receive on the input frequency.

> QUESTION: What is the most common use of the "reverse split" function of a VHF/UHF transceiver? (T2B01)
>
> ANSWER: **Listen on a repeater's input frequency**

Another problem you may encounter is over-deviation. This can happen if you speak too loudly into the microphone. This will cause your signal to deviate too much, and that can cause distortion.

> QUESTION: What might be the problem if a repeater user says your transmissions are breaking up on voice peaks? (T2B05)
>
> ANSWER: **You are talking too loudly**

In addition to knowing how to set the controls of your radio, you need to know the protocol for making contacts. When using a repeater, the protocol is very simple. The reason for this is that signals are normally very strong and saying your call sign is all that is required to alert other stations that you are listening to the repeater and available for contacts.

> QUESTION: What is an appropriate way to call another station on a repeater if you know the other station's call sign? (T2A04)
>
> ANSWER: **Say the station's call sign, then identify with your call sign**
>
> QUESTION: What brief statement indicates that you are listening on a repeater and looking for a contact? (T2A09)
>
> ANSWER: **Your call sign**

Repeater operation is called duplex operation because you're transmitting and receiving on two different frequencies. When two stations are operating on the same frequency, without the aid of a repeater, it's called simplex operation. On the 2 m band, simplex channels start at 146.52 MHz. On the 70 cm band, simplex channels start at 446.00 MHz.

> QUESTION: What term describes an amateur station that is transmitting and receiving on the same frequency? (T2A11)
>
> ANSWER: **Simplex**
>
> QUESTION: Why are simplex channels designated in the VHF/UHF band plans? (T2B12)
>
> ANSWER: **So that stations within mutual communications range can communicate without tying up a repeater**

To help amateurs operating simplex find one another, frequencies on each band have been set aside as "national calling frequencies." 146.52 MHz is the national calling frequency for FM simplex operation in the 2 m band. 446.000 MHz is the national calling frequency for the 70 cm band.

QUESTION: What is the national calling frequency for FM simplex operations in the 2 meter band? (T2A02)

ANSWER: 146.520 MHz

HF Operation

On the HF bands, signals can be easy to copy or difficult to copy. Because this is the case, the protocol for making contacts is more complex than the repeater protocol. On HF, when you want to contact another station, you "call CQ." That is to say, you would say something like, "CQ CQ CQ. This is KB6NU." This means that you are open to a call from any station.

> QUESTION: What is the meaning of the procedural signal "CQ"? (T2A08)
>
> ANSWER: **Calling any station**

You don't want to just start calling CQ willy-nilly, though.

> QUESTION: Which of the following is a guideline when choosing an operating frequency for calling CQ? (T2A12)
>
> ANSWER: **All of these choices are correct**
>
> - Listen first to be sure that no one else is using the frequency
> - Ask if the frequency is in use
> - Make sure you are in your assigned band

Knowing how to reply to a CQ is also important. Knowing the commonly accepted protocol will make it easier to make contacts.

> QUESTION: How should you respond to a station calling CQ? (T2A05)
>
> ANSWER: **Transmit the other station's call sign followed by your call sign**

For example, if W8JNZ heard my call and wanted to talk to me, he would reply, "KB6NU this is W8JNZ. Over." Then, I would return the call, and our contact would begin. If signal conditions are poor, you may want to repeat your call sign and state your call sign in a phonetic alphabet.

It's important to always identify your station, even when only performing tests.

> QUESTION: Which of the following is required when making on-the-air test transmissions? (T2A06)
>
> ANSWER: **Identify the transmitting station**

As a Technician, you will be able to operate Morse Code on certain portions of the 80 m, 40 m, 15 m, and 10 m bands. To shorten the number of characters sent during a CW contact, amateurs often use three-letter combinations called Q-signals. Q signals are three-letter combinations, beginning with the letter "Q," that stand for commonly-used phrases. You need to know the meaning of two of these Q signals: QRM and QSY.

> QUESTION: Which Q signal indicates that you are receiving interference from other stations? (T2B10)
>
> ANSWER: **QRM**
>
> QUESTION: Which Q signal indicates that you are changing frequency? (T2B11)
>
> ANSWER: **QSY**

FCC rules specify broadly where amateur radio operators have operating privileges, but they are not very detailed. Band plans take this one step further, suggesting where amateurs should use certain modes.

> QUESTION: What is a band plan, beyond the privileges established by the FCC? (T2A10)
>
> ANSWER: **A voluntary guideline for using different modes or activities within an amateur band**

While not always adhered to, another basic tenet of amateur radio is to operate courteously and avoid interfering with other stations.

QUESTION: Which of the following applies when two stations transmitting on the same frequency interfere with each other? (T2B08)

ANSWER: **Common courtesy should prevail, but no one has absolute right to an amateur frequency**

Public service and emergency communications

One of the reasons amateur radio exists at all is that ham radio operators are uniquely set up to provide emergency and public-service communications. As a result, many hams consider it an obligation to be prepared to help out when called upon to do so. This includes having the proper equipment and knowing the proper operating procedures. There are two organizations that provide emergency communications: the Radio Amateur Civil Emergency Service (RACES) and the Amateur Radio Emergency Service (ARES).

QUESTION: What is the Amateur Radio Emergency Service (ARES)? (T2C12)

ANSWER: **Licensed amateurs who have voluntarily registered their qualifications and equipment for communications duty in the public service**

QUESTION: Which of the following describes the Radio Amateur Civil Emergency Service (RACES)? (T1A10)

ANSWER: **All of these choices are correct**

- A radio service using amateur frequencies for emergency management or civil defense communications

- A radio service using amateur stations for emergency management or civil defense communications

- An emergency service using amateur operators certified by a civil defense organization as being enrolled in that organization

QUESTION: What do RACES and ARES have in common? (T2C04)

ANSWER: **Both organizations may provide communications during emergencies**

When an emergency occurs, it's common for amateur radio operators to form a network or "net" to facilitate emergency communications. The net is led by the net control station, or NCS, whose job it is to make sure that messages are passed in an efficient and timely manner. Stations other than the net control station are said to "check into" the net.

QUESTION: What is meant by the term "NCS" used in net operation? (T2C02)

ANSWER: **Net Control Station**

QUESTION: Which of the following is an accepted practice for an amateur operator who has checked into a net? (T2C07)

ANSWER: **Remain on frequency without transmitting until asked to do so by the net control station**

There are, however, times when a station may need to get the immediate attention of the net control station.

QUESTION: Which of the following is an accepted practice to get the immediate attention of a net control station when reporting an emergency? (T2C06)

ANSWER: **Begin your transmission by saying "Priority" or "Emergency" followed by your call sign**

The term for messages passed between stations in a net is "traffic," and the process of passing messages to and from amateur radio stations is called "handling traffic."

QUESTION: What does the term "traffic" refer to in net operation? (T2C05)

ANSWER: **Formal messages exchanged by net stations**

QUESTION: Which of the following is a characteristic of good traffic handling? (T2C08)

ANSWER: **Passing messages exactly as received**

QUESTION: What should be done when using voice modes to ensure that voice messages containing unusual words are received correctly? (T2C03)

ANSWER: **Spell the words using a standard phonetic alphabet**

Formal traffic messages consists of four parts: preamble, address, text, signature. Part of the preamble is the check. The address is the name and address of the intended recipient, the text is the message itself, and the signature is the part of the message that identifies the originator of the message.

QUESTION: What information is contained in the preamble of a formal traffic message? (T2C10)

ANSWER: **The information needed to track the message**

QUESTION: What is meant by the term "check," in reference to a formal traffic message? (T2C11)

ANSWER: **The number of words or word equivalents in the text portion of the message**

Even in emergencies, you must follow FCC rules when operating an amateur radio station. The normal rules are, however, relaxed a little during true emergencies.

QUESTION: When do the FCC rules NOT apply to the operation of an amateur station? (T2C01)

ANSWER: **FCC rules always apply**

QUESTION: Are amateur station control operators ever permitted to operate outside the frequency privileges of their license class? (T2C09)

ANSWER: **Yes, but only if necessary in situations involving the immediate safety of human life or protection of property**

It's kind of a Catch-22. FCC rules always apply to the operation of an amateur radio station, but the rules say that you can do almost anything in a true emergency.

Amateur satellite operation

Making contacts via amateur radio satellites and other space stations is one of the coolest things a ham can do. As a Technician Class licensee, you will have the privileges to do this.

> QUESTION: Which amateur radio stations may make contact with an amateur radio station on the International Space Station (ISS) using 2 meter and 70 cm band frequencies? (T1B02)
>
> ANSWER: **Any amateur holding a Technician or higher-class license**

Amateur satellites are basically repeaters in space. As such they have an uplink frequency, which is the frequency on which you transmit and the satellite receives, and a downlink frequency, on which the satellite transmits and you receive. Often, the uplink frequency and downlink frequency are in different amateur bands.

> QUESTION: What is meant by the statement that a satellite is operating in mode U/V? (T8B08)
>
> ANSWER: **The satellite uplink is in the 70 cm band and the downlink is in the 2 meter band**

The 70 cm band is in the UHF portion of the spectrum, hence the "U" in U/V, while the 2 meter band is in the VHF portion of the spectrum, hence the "V" in U/V.

While most satellites are FM satellites, some operate using other modes.

> QUESTION: What mode of transmission is commonly used by amateur radio satellites? (T8B04)
>
> ANSWER: **All of these choices are correct**
>
> - SSB

- FM
- CW/data

When making contacts via an amateur satellite only use as much power as is needed to make the contact. The reason for this is that when a satellite receives a very strong signal, its automatic gain control (AGC) sets the receive threshold to the level of that signal and weaker signals won't be relayed. When everyone uses a reasonable power level, the AGC doesn't kick in, and the satellite can relay many signals simultaneously.

QUESTION: What is the impact of using too much effective radiated power on a satellite uplink? (T8B02)

ANSWER: **Blocking access by other users**

QUESTION: Which of the following is a good way to judge whether your uplink power is neither too low nor too high? (T8B12)

ANSWER: **Your signal strength on the downlink should be about the same as the beacon**

Most amateur satellites are in a low Earth orbit, or LEO. Satellites in a low Earth orbit have an altitude between 99 miles and 1,200 miles. This corresponds to an orbital period of about 88 minutes to about about 127 minutes. Satellites in LEO provides high bandwidth and low communication time lag, but they can only be used for a short time when they pass overhead.

QUESTION: What do the initials LEO tell you about an amateur satellite? (T8B10)

ANSWER: **The satellite is in a Low Earth Orbit**

Amateur satellites are often equipped with beacons. Beacons often send telemetry signals that inform users about the status of the satellite.

QUESTION: What is a satellite beacon? (T8B05)

ANSWER: **A transmission from a satellite that contains status information**

QUESTION: What telemetry information is typically transmitted by satellite beacons? (T8B01)

ANSWER: **Health and status of the satellite**

QUESTION: Who may receive telemetry from a space station? (T8B11)

ANSWER: **Anyone who can receive the telemetry signal**

Computers make it easy to figure out when you can communicate via an amateur satellite. Computer programs are available that not only tell you when a satellite is passing overhead, but also control an antenna rotor and set the frequency of your transceiver.

QUESTION: Which of the following are provided by satellite tracking programs? (T8B03)

ANSWER: **All of these answers are correct**

- Maps showing the real-time position of the satellite track over the earth
- The time, azimuth, and elevation of the start, maximum altitude, and end of a pass
- The apparent frequency of the satellite transmission, including effects of Doppler shift

QUESTION: Which of the following are inputs to a satellite tracking program? (T8B06)

ANSWER: **The Keplerian elements**

Two issues that you must deal with when communicating via satellites are Doppler shift and spin fading.

QUESTION: With regard to satellite communications, what is Doppler shift? (T8B07)

ANSWER: **An observed change in signal frequency caused by relative motion between the satellite and the earth station**

QUESTION: What causes "spin fading" of satellite signals? (T8B09)

ANSWER: **Rotation of the satellite and its antennas**

Operating activities

There are many different ways to have fun with amateur radio. Contesting is one of them.

> QUESTION: What operating activity involves contacting as many stations as possible during a specified period? (T8C03)
>
> ANSWER: **Contesting**
>
> QUESTION: Which of the following is a good procedure when contacting another station in a radio contest? (T8C04)
>
> ANSWER: **Send only the minimum information needed for proper identification and the contest exchange**

Sending the minimum amount of information will help you make as many contacts as possible.

Information about a station's location is often part of the contest exchange. In the U.S., a station's state or ARRL section is most often sent, but in VHF/UHF contests, stations often send each other their grid locators.

> QUESTION: What is a grid locator? (T8C05)
>
> ANSWER: **A letter-number designator assigned to a geographic location**

One activity that is both fun and practical is radio direction finding. You use radio direction finding equipment and skills to participate in hidden transmitter hunts.

> QUESTION: Which of the following methods is used to locate sources of noise interference or jamming? (T8C01)
>
> ANSWER: **Radio direction finding**
>
> QUESTION: Which of these items would be useful for a hidden transmitter hunt? (T8C02)

ANSWER: **A directional antenna**

If the only radios that you have are VHF or UHF radios, you might want to look into EchoLink and the Internet Radio Linking Project (IRLP). Both systems provide a way to communicate with amateurs far away with a VHF or UHF transceiver. Both use Voice Over Internet Protocol (VoIP).

QUESTION: What is the Internet Radio Linking Project (IRLP)? (T8C08)

ANSWER: **A technique to connect amateur radio systems, such as repeaters, via the internet using Voice Over Internet Protocol (VoIP)**

QUESTION: What is meant by Voice Over Internet Protocol (VoIP), as used in amateur radio? (T8C07)

ANSWER: **A method of delivering voice communications over the internet using digital techniques**

QUESTION: What must be done before you may use the EchoLink system to communicate using a repeater? (T8C10)

ANSWER: **You must register your call sign and provide proof of license**

Stations that connect to EchoLink or IRLP are called nodes.

QUESTION: How might you obtain a list of active nodes that use VoIP? (T8C09)

ANSWER: **All of these choices are correct**

- By subscribing to an on line service
- From on line repeater lists maintained by the local repeater frequency coordinator
- From a repeater directory

QUESTION: How is access to some IRLP nodes accomplished? (T8C06)

ANSWER: **By using DTMF signals**

QUESTION: What type of tones are used to control repeaters linked by the Internet Relay Linking Project (IRLP) protocol? (T2B06)

ANSWER: **DTMF**

Sometimes, nodes are also gateways.

QUESTION: What name is given to an amateur radio station that is used to connect other amateur stations to the internet? (T8C11)

ANSWER: **A gateway**

Rules and regulations

Purpose and permissible use of the Amateur Radio Service, operator/primary station license grant; basic terms used in FCC rules; interference; RACES rules; phonetics; Frequency Coordinator

The Amateur Radio Service is a service administered by the Federal Communications Commission (FCC). The FCC establishes the rules and regulations which govern the service.

> QUESTION: Which agency regulates and enforces the rules for the Amateur Radio Service in the United States? (T1A02)
>
> ANSWER: **The FCC**

Part 97 is the part of the radio regulations that govern the Amateur Radio Service. Part 97.1 lists five "purposes" for the existence of amateur radio. The first is recognition of its usefulness in providing emergency and public-service communications. Another is the use of amateur radio as a way to help people become better technicians and operators.

> QUESTION: Which of the following is a purpose of the Amateur Radio Service as stated in the FCC rules and regulations? (T1A01)

ANSWER: **Advancing skills in the technical and communication phases of the radio art**

Part 97 defines terms and concepts that every amateur radio operator needs to know.

> QUESTION: What is the FCC Part 97 definition of a beacon? (T1A06)

> ANSWER: **An amateur station transmitting communications for the purposes of observing propagation or related experimental activities**

> QUESTION: What is the FCC Part 97 definition of a space station? (T1A07)

> ANSWER: **An amateur station located more than 50 km above the Earth's surface**

One of the most important concepts in Part 97 is that of harmful interference. Part 97 defines harmful interference as "interference which endangers the functioning of a radionavigation service or of other safety services or seriously degrades, obstructs or repeatedly interrupts a radiocommunication service operating in accordance with the Radio Regulations."

> QUESTION: When is willful interference to other amateur radio stations permitted? (T1A11)

> ANSWER: **At no time**

Part 97 also contains rules about how repeater frequencies are assigned.

> QUESTION: Which of the following entities recommends transmit/receive channels and other parameters for auxiliary and repeater stations? (T1A08)

ANSWER: **Volunteer Frequency Coordinator recognized by local amateurs**

QUESTION: Who selects a Frequency Coordinator? (T1A09)

ANSWER: **Amateur operators in a local or regional area whose stations are eligible to be repeater or auxiliary stations**

Authorized frequencies: frequency allocations; ITU; emission modes; restricted sub-bands; spectrum sharing; transmissions near band edges; contacting the International Space Station; power output

The International Telecommunications Union (ITU) is the body responsible for setting international telecommunications rules and regulations. This includes amateur radio.

QUESTION: What is the International Telecommunications Union (ITU)? (T1B01)

ANSWER: **A United Nations agency for information and communication technology issues**

Because operation outside of the amateur radio bands is a serious offense, it is important to know about the frequencies that amateur radio operators can use, as well as the modes you can use on those frequencies.

QUESTION: Which frequency is within the 6 meter amateur band? (T1B03)

ANSWER: **52.525 MHz**

QUESTION: Which amateur band are you using when your station is transmitting on 146.52 MHz? (T1B04)

ANSWER: **2 meter band**

QUESTION: Why should you not set your transmit frequency to be exactly at the edge of an amateur band or sub-band? (T1B09)

ANSWER: **All of these choices are correct**

- To allow for calibration error in the transmitter frequency display
- So that modulation sidebands do not extend beyond the band

edge

- To allow for transmitter frequency drift

QUESTION: What is the limitation for emissions on the frequencies between 219 and 220 MHz? (T1B05)

ANSWER: **Fixed digital message forwarding systems only**

QUESTION: On which HF bands does a Technician class operator have phone privileges? (T1B06)

ANSWER: **10 meter band only**

QUESTION: Which of the following HF bands have frequencies available to the Technician class operator for RTTY and data transmissions? (T1B10)

ANSWER: **10 meter band only**

QUESTION: Which of the following VHF/UHF frequency ranges are limited to CW only? (T1B07)

ANSWER: **50.0 MHz to 50.1 MHz and 144.0 MHz to 144.1 MHz**

Since Technician Class operators have full amateur privileges above 50 MHz, they can operate transmitters with an output power of up to 1,500 watts at frequencies in the VHF region and above. On the HF bands, however, transmitters operated by Technicians are restricted to an output power of 200 watts or less.

QUESTION: What is the maximum peak envelope power output for Technician class operators using their assigned portions of the HF bands? (T1B11)

ANSWER: **200 watts**

QUESTION: Except for some specific restrictions, what is the maximum peak envelope power output for Technician class operators using frequencies above 30 MHz? (T1B12)

ANSWER: **1500 watts**

Amateur radio operators share some bands with users from other services. Sometimes, amateurs are the primary users, such as in the 2m band, but sometimes amateur radio operators are secondary users.

QUESTION: Which of the following is a result of the fact that the Amateur Radio Service is secondary in all or portions of some amateur bands (such as portions of the 70 cm band)? (T1B08)

ANSWER: **U.S. amateurs may find non-amateur stations in those portions, and must avoid interfering with them**

Operator licensing: operator classes; sequential and vanity call sign systems; international communications; reciprocal operation; places where the Amateur Radio Service is regulated by the FCC; name and address on FCC license database; license term; renewal; grace period

As you might expect, licensing is a big deal in the Amateur Radio Service. Your class of license determines where you can operate, and in some cases, what modes you can operate and how much power you can use.

QUESTION: For which license classes are new licenses currently available from the FCC? (T1C01)

ANSWER: **Technician, General, Amateur Extra**

QUESTION: What is the normal term for an FCC-issued primary station/operator amateur radio license grant? (T1C08)

ANSWER: **Ten years**

QUESTION: How soon after passing the examination for your first amateur radio license may you operate a transmitter on an Amateur Radio Service frequency? (T1C10)

ANSWER: **As soon as your operator/station license grant appears in the FCC's license database**

For some time now, the official amateur radio license authorization has been the electronic record that exists in the FCC Universal Licensing System (ULS). Paper licenses are no longer issued as a matter of course, although you can log into the FCC website and print out a paper copy if you so choose.

QUESTION: What is proof of possession of an FCC-issued operator/primary license grant? (T1A05)

ANSWER: **The control operator's operator/primary station**

license must appear in the FCC ULS consolidated licensee database

After you pass the test, the FCC will assign you a call sign sequentially from the pool of available call signs. If you do not like this call sign, you can apply for a vanity call sign.

> QUESTION: Who may select a desired call sign under the vanity call sign rules? (T1C02)
>
> ANSWER: **Any licensed amateur**

The call sign you select must not only be available, it must have an appropriate format for the class of license you hold. For example, only Amateur Extra class licensees may hold 1x2 or 2x1 call signs. This means that a Technician class amateur radio operator may not choose the call signs KA1X, which is a 2x1 call sign, or W1XX, which is a 1x2 call sign.

> QUESTION: Which of the following is a valid call sign for a Technician class amateur radio station? (T1C05)
>
> ANSWER: **K1XXX**

If you don't renew your license before it expires, or within the two-year grace period, you will have to take the test again to get a new amateur radio license.

> QUESTION: What is the grace period following the expiration of an amateur license within which the license may be renewed? (T1C09)
>
> ANSWER: **Two years**
>
> QUESTION: If your license has expired and is still within the allowable grace period, may you continue to operate a transmitter on Amateur Radio Service frequencies? (T1C11)

ANSWER: **No, transmitting is not allowed until the FCC license database shows that the license has been renewed**

Clubs may apply for a station license for their club station. The club may even apply for a vanity call sign.

QUESTION: Which of the following is a requirement for the issuance of a club station license grant? (T1F11)

ANSWER: **The club must have at least four members**

When you get your first license, you must give the examiners a mailing address. Should you move, you must inform the FCC of your new mailing address.

QUESTION: What may result when correspondence from the FCC is returned as undeliverable because the grantee failed to provide and maintain a correct mailing address with the FCC? (T1C07)

ANSWER: **Revocation of the station license or suspension of the operator license**

Some countries have reciprocal licensing agreements with the U.S., and you can operate from that country without any specific authorization. For example, I could operate my station in Germany by simply using the call sign DL/KB6NU. There are restrictions on your operating privileges, depending on the country from which you plan to operate, and you should investigate these before you get on the air.

QUESTION: When are you allowed to operate your amateur station in a foreign country? (T1C04)

ANSWER: **When the foreign country authorizes it**

You can also operate your station while aboard a ship in international waters.

QUESTION: From which of the following locations may an FCC-licensed amateur station transmit? (T1C06)

ANSWER: **From any vessel or craft located in international waters and documented or registered in the United States**

Authorized and prohibited transmission: communications with other countries; music; exchange of information with other services; indecent language; compensation for use of station; retransmission of other amateur signals; codes and ciphers; sale of equipment; unidentified transmissions; one-way transmission

As a licensed radio amateur, it's important to know what you can and can't do on the air. Indecent language is prohibited, and oddly enough, so is music, except for one specific situation.

> QUESTION: What, if any, are the restrictions concerning transmission of language that may be considered obscene or indecent? (T1D06)
>
> ANSWER: **Any such language is prohibited**
>
> QUESTION: Under what conditions is an amateur station authorized to transmit music using a phone emission? (T1D04)
>
> ANSWER: **When incidental to an authorized retransmission of manned spacecraft communications**

Transmitting any codes whose specifications are not published or well-known is prohibited, except in one specific circumstance.

> QUESTION: When is it permissible to transmit messages encoded to hide their meaning? (T1D03)
>
> ANSWER: **Only when transmitting control commands to space stations or radio control craft**

Amateur radio stations may only communicate with amateur stations in other countries when that country allows it.

> QUESTION: With which countries are FCC-licensed amateur radio stations prohibited from exchanging

communications? (T1D01)

ANSWER: **Any country whose administration has notified the International Telecommunications Union (ITU) that it objects to such communications**

Currently, there are no countries that U.S. amateurs are prohibited from contacting.

Another big deal in amateur radio is the prohibition of being paid to operate an amateur radio station, except in some very special circumstances. That doesn't mean that you can't make money from amateur radio. I'm obviously making a few bucks by selling study guides, but I can't be paid for operating my station or someone else's station.

QUESTION: In which of the following circumstances may the control operator of an amateur station receive compensation for operating that station? (T1D08)

ANSWER: **When the communication is incidental to classroom instruction at an educational institution**

QUESTION: When may amateur radio operators use their stations to notify other amateurs of the availability of equipment for sale or trade? (T1D05)

ANSWER: **When the equipment is normally used in an amateur station and such activity is not conducted on a regular basis**

All amateur communications must be station to station. That is to say, amateur radio operators may not broadcast.

QUESTION: What is the meaning of the term broadcasting in the FCC rules for the Amateur Radio Service? (T1D10).

ANSWER: **Transmissions intended for reception by the general public**

QUESTION: Under which of the following circumstances are amateur stations authorized to transmit signals related to broadcasting, program production, or news gathering, assuming no other means is available? (T1D09)

ANSWER: **Only where such communications directly relate to the immediate safety of human life or protection of property**

As with many rules, however, there are exceptions.

QUESTION: Under which of the following circumstances may an amateur radio station make one-way transmissions? (T1D02)

ANSWER: **When transmitting code practice, information bulletins, or transmissions necessary to provide emergency communications**

So, what is allowed?

QUESTION: What types of international communications is an FCC-licensed amateur radio station permitted to make? (T1C03)

ANSWER: **Communications incidental to the purposes of the Amateur Radio Service and remarks of a personal character**

Control operator and control types: control operator required; eligibility; designation of control operator; privileges and duties; control point; local, automatic and remote control; location of control operator

An important concept in amateur radio is the control operator. The basic concept is that an amateur radio station must always have a control operator, and that control operator is responsible for the proper operation of that station. And, the default control operator is the station licensee.

QUESTION: When is an amateur station permitted to transmit without a control operator? (T1E01)

ANSWER: **Never**

QUESTION: Who does the FCC presume to be the control operator of an amateur station, unless documentation to the contrary is in the station records? (T1E11)

ANSWER: **The station licensee**

QUESTION: Who must designate the station control operator? (T1E03)

ANSWER: **The station licensee**

QUESTION: When the control operator is not the station licensee, who is responsible for the proper operation of the station? (T1E07)

ANSWER: **The control operator and the station licensee are equally responsible**

QUESTION: Who is accountable should a repeater inadvertently retransmit communications that violate the FCC rules? (T1F10)

ANSWER: **The control operator of the originating station**

QUESTION: What determines the transmitting privileges of an amateur station? (T1E04)

ANSWER: **The class of operator license held by the control operator**

QUESTION: When, under normal circumstances, may a Technician class licensee be the control operator of a station operating in an exclusive Amateur Extra class operator segment of the amateur bands? (T1E06)

ANSWER: **At no time**

QUESTION: Who may be the control operator of a station communicating through an amateur satellite or space station? (T1E02)

ANSWER: **Any amateur whose license privileges allow them to transmit on the satellite uplink frequency**

Two related concepts are the control point and control type. Part 97 defines three control types:
- Local control. A station is said to be locally controlled when the control operator can directly manipulate the operating of an amateur radio station.
- Remote control. A station is said to be remotely controlled when the control operator indirectly manipulates the operating controls of an amateur radio station through a control link, such as a radio link, a telephone link, or an internet link.
- Automatic control. A station is said to be automatically controlled if it uses devices and procedures for control without the control operator being present at the control point.

QUESTION: What is an amateur station control point? (T1E05)

ANSWER: **The location at which the control operator function is performed**

QUESTION: Which of the following is an example of automatic control? (T1E08)

ANSWER: **Repeater operation**

QUESTION: Which of the following is an example of remote control as defined in Part 97? (T1E10)

ANSWER: **Operating the station over the internet**

QUESTION: Which of the following is true of remote control operation? (T1E09)

ANSWER: **All of these choices are correct**

- The control operator must be at the control point
- A control operator is required at all times
- The control operator indirectly manipulates the controls

Station identification, repeaters, third-party communications, FCC inspection

Proper station identification is also very important. In fact, failure to identify properly is perhaps the most common rule violation.

QUESTION: When is an amateur station required to transmit its assigned call sign? (T1F03)

ANSWER: **At least every 10 minutes during and at the end of a communication**

QUESTION: When may an amateur station transmit without on-the-air identification? (T1D11)

ANSWER: **When transmitting signals to control a model craft**

QUESTION: Which of the following is an acceptable language to use for station identification when operating in a phone sub-band? (T1F04)

ANSWER: **The English language**

QUESTION: What method of call sign identification is required for a station transmitting phone signals? (T1F05)

ANSWER: **Send the call sign using a CW or phone emission**

QUESTION: What are the FCC rules regarding the use of a phonetic alphabet for station identification in the Amateur Radio Service? (T1A03)

ANSWER: **It is encouraged**

For some types of operations, using a tactical call is allowed. A tactical call describes the function of the station or the location of a station.

QUESTION: When using tactical identifiers such as "Race Headquarters" during a community service net operation, how often must your station transmit the station's FCC-assigned call sign? (T1F02)

ANSWER: **At the end of each communication and every ten minutes during a communication**

When operating mobile or portable, or when you wish to note something about your station, you may use a self-assigned call sign indicator, such as "/3," "mobile," or "QRP."

QUESTION: Which of the following formats of a self-assigned indicator is acceptable when identifying using a phone transmission? (T1F06)

ANSWER: **All of these choices are correct**

- KL7CC stroke W3
- KL7CC slant W3
- KL7CC slash W3

Third-party communications are communications on behalf of someone who is not the station licensee. For example, if you have a friend over to your house and let him or her talk on your radio, that is a third-party communication. These are entirely legal within the United States, but there are some restrictions when you are in contact with an amateur station in a foreign country.

QUESTION: What is meant by the term Third Party Communications? (T1F08)

ANSWER: **A message from a control operator to another amateur station control operator on behalf of another person**

QUESTION: Which of the following restrictions apply when a non-licensed person is allowed to speak to a foreign station using a station under the control of a Technician class control operator? (T1F07)

ANSWER: **The foreign station must be one with which the U.S. has a third-party agreement**

Finally—and I do mean finally.

QUESTION: When must the station licensee make the station and its records available for FCC inspection? (T1F01)

ANSWER: **At any time upon request by an FCC representative**

They're not going to knock on your door at 3 a.m. some morning to take a look at your shack, but one of your obligations as a licensee is to make your station and your records available when requested to do so.

Well, that's it! We've covered all 424 questions in the Technician Class question pool. Now, you should take some online practice tests, and when you're passing those regularly, find an exam session and get your license. Good luck and 73!

DAN ROMANCHIK, KB6NU

Glossary

AC: alternating current. Alternating current is the name for current that reverses direction on a regular basis. (T5A09). The power outlets in your home provide alternating current.

APRS: Automatic Packet Reporting System. APRS is digital communications system used by amateur radio operators. While it is normally used for tracking the location of mobile stations, it can be used for other purposes as well. For more information, go to http://www.aprs.org.

ARES: Amateur Radio Emergency Service. The Amateur Radio Emergency Service consists of licensed amateurs who have voluntarily registered their qualifications and equipment with their local ARES leadership for communications duty in the public service when disaster strikes. For more information, go to http://www.arrl.org/ares.

AM: amplitude modulation. The type of modulation that varies the amplitude of a radio signal in accordance with the amplitude of a modulating signal. For more information, go to http://www.pa2old.nl/files/am_fundamentals.pdf.

CTCSS: Continuous Tone Coded Squelch System. A system that uses sub-audible tones, transmitted along with the audio portion of a transmission to control whether or not a repeater will re-transmit a signal. It is known by a number of different trade names, including Private Line® (PL) by Motorola. In practice, it's used to prevent nearby transmitters from inadvertently turning on repeaters.

CW: continuous wave. This is the operating mode amateur radio operators use when sending Morse Code.

DC: direct current. Direct current is the name for current that never reverses direction.

DMR: Digital Mobile Radio

DTMF: dual-tone, multi-frequency. DTMF is a type of signaling used to send data over voice channels. Its most common use in amateur radio is to allow users of handheld transceivers to send commands to repeater systems. It is called DTMF because every time a user presses a keypad button a unique tone consisting of two frequencies is transmitted. For more information, see http://www.genave.com/dtmf.htm.

FCC: Federal Communications Commission. This is the government body which sets the rules for amateur radio in the U.S.

FM: frequency modulation. The type of modulation normally used when operating on VHF and UHF repeaters.

HF: high frequency. The range of frequencies between 3 MHz and 30 MHz.

HT: handy-talky or handheld transceiver. "Handy Talky" is a Motorola trademark.

ITU: International Telecommunications Union. This is the international body which governs amateur radio worldwide.

LSB: lower sideband. See **SSB**.

MFSK: multi-frequency shift keying. A type of modulation used to send digital information over a radio channel.

PL: Private Line. See **CTCSS**.

PSK: phase shift keying. A method for sending digital information over a radio channel. A popular amateur radio "digital mode" is PSK31, which uses PSK modulation and occupies only 31 Hz of bandwidth.

PTT: push-to-talk

RACES: Radio Amateur Civil Emergency Service. RACES is an amateur radio emergency communications service created by the Federal Emergency Management Agency (FEMA) and the FCC. RACES volunteers serve their respective jurisdictions pursuant to guidelines and mandates established by local emergency management officials. See http://www.usraces.org/ for more information.

RIT: receiver incremental tuning. A control which allows a user to set the receive frequency of a transceiver either slightly higher or slightly lower than the transmit frequency.

RF: radio frequency

SSB: single sideband. When a carrier is amplitude modulated, both upper and lower sidebands are produced. This results in a signal that is 6 kHz wide. Since both sidebands carry the same information, and the carrier carries no information, someone figured out that if they could filter out the carrier and one of the sidebands, and put all the power into a single sideband, the efficiency of voice communications would be much greater. Nearly all voice communications on the shortwave bands now use SSB.

SWR: standing-wave ratio. The SWR of an antenna system is a measure of how closely the impedances of the antenna and feedline match the output impedance of the transmitter.

VHF: very high frequency. The range of frequencies between 30 MHz and 300 MHz.

ULS: Universal Licensing System. The FCC's Universal Licensing system contains information on all FCC licensees, including amateur radio operators. For more information, go to http://www.fcc.gov/uls.

UHF: ultra high frequency. The range of frequencies between 300 MHz and 3000 MHz. 41

USB: upper sideband. See **SSB**.

VFO: variable frequency oscillator. VFOs are used to control the receiving and transmitting frequencies of amateur radio equipment.

DAN ROMANCHIK, KB6NU

About the Author

I have been a ham radio operator since 1971 and a radio enthusiast as long as I can remember. I've been teaching ham radio classes for the past twelve years. In addition to being an amateur radio instructor:

- I blog about amateur radio at KB6NU.Com (http://kb6nu.com).
- I have written study guides for the General Class and Extra Class exams. You can find the *No-Nonsense General Class License Study Guide* and the *No-Nonsense Extra Class License Study Guide* in PDF, Nook (ePub) and Kindle (.mobi) formats on my website at http://www.kb6nu.com/study-guides. It's also available in print from Amazon.
- I am the author of *21 Things to Do With your Amateur Radio License*, a book for those who have been recently licensed or are just getting back into the hobby. It's also available in PDF, Nook (ePub) and Kindle (.mobi) formats on my website at http://www.kb6nu.com/product/21-things-to-do-after-you-get-your-amateur-radio-license-pdf/. It's also available in print from Amazon.
- I am the author of *The CW Geek's Guide to Having Fun with Morse Code*, a book for those wanting to learn the art of Morse Code. It's also available in PDF, Nook (ePub) and Kindle (.mobi) formats on my website at https://www.kb6nu.com/product/cw-geeks-guide-to-having-fun-with-morse-code-pdf/. It's also available in print from Amazon.
- I teach amateur radio classes all over the country, including an annual class at the Dayton Hamvention.

You can contact me by sending e-mail to cwgeek@kb6nu.com. If you have comments or questions about any of the stuff in this book, I hope you will do so.

73!
Dan, KB6NU

Made in the USA
Middletown, DE
11 February 2020